COUNTING
SHEEP

A Celebration of the Pastoral Heritage of Britain

PHILIP WALLING

PROFILE BOOKS

First published in Great Britain in 2014 by
PROFILE BOOKS LTD
3A Exmouth House
Pine Street
London ECIR OJH
www.profilebooks.com

1 3 5 7 9 10 8 6 4 2

Typeset in Dante by MacGuru Ltd
info@macguru.org.uk
Printed and bound in Great Britain by
Clays, Bungay, Suffolk

A CIP catalogue record for this book is available from the
British Library.

ISBN 978 1 84668 504 0
eISBN 978 1 84765 803 6

The paper this book is printed on is certified by the © 1996 Forest Stewardship
Council A.C. (FSC). It is ancient-forest friendly.
The printer holds FSC chain of custody SGS-COC-2061

For Stephanie, who believed in me
because I didn't believe in myself.

❧ CONTENTS ❧

ᴠᴠ ACKNOWLEDGEMENTS ᴠᴠ

I N A HOWLING BLIZZARD DURING THE WINTER OF 1940–41 my grandfather received a phone call from Cockermouth asking him to form a search party to look for a young soldier called Sale, who had left the farm where he had been on leave with his aunt and her family to walk the six miles to Cockermouth railway station to rejoin his regiment, but had not arrived. His family had evacuated themselves for the duration of the war from Aylesbury to Picket How, their farm between Lorton and Loweswater. The searchers luckily found the soldier in a snow drift, exhausted and insensible and huddled in his greatcoat. They took him to the district nurse's house, where he was revived. Ever afterwards, the Sale family maintained that my grandfather's search party had saved their nephew's life.

Thirty-five years later, when I was looking for a farm to rent, the Sales had not forgotten the events of that terrible winter. Their tenants at Picket How, the Mackereth brothers, were retiring after a hundred years on the farm and the Sales offered me the tenancy at a very low rent. I did not enquire why the rent was little more than nominal, nor did I know at that time about my grandfather's search party. I just accepted the generosity that gave me such a flying start in farming. Later, when they decided to sell the farm, I was able to buy it at a substantial discount as sitting tenant. When I left farming to go into the law, and subsequently sold the farm, I felt a terrible sense of guilt that I appeared to have thrown their kindness back in their faces, especially because I was sure they believed I would live there for the rest of my life.

I owe a deep debt of gratitude to the Sales for their almost transcendent generosity to a young man, at the beginning of his life, when he needed it most and I hope this book will be accepted as slight recompense, inadequate as it is.

It is hard to know exactly which influences and people contribute to the writing of a book, but I know I have incurred many unrepayable debts to many people. For without their willingness to indulge me, often for hours on end, I would never have been able to start let alone finish it. Most of the farmers and breeders I simply telephoned out of the blue and, not knowing me from Adam, they invited me to their farms, showed me round, patiently answered my questions and told me more about themselves, their sheep and their farming lives than I was entitled to know. Everywhere I went I was received with considerable courtesy and hospitality, not to say kindness, and it is a pleasant duty to record this. If I have inadvertently omitted anyone to whom I owe thanks, I can only apologise and ask for their forgiveness.

My first point of contact was usually with the secretary of each breed society and I wish to thank them all for their helpfulness in usually directing me to a breeder with the patience to deal with me. I thank them all for their courtesy and patience.

For half a day I shamelessly picked the brains of John Thorley MBE, for many years the Chief Executive of the National Sheep Association, now the power behind the Pastoral Alliance and an indefatigable pillar of the Campaign for Wool. I am immensely grateful to him for sharing with me his fathomless knowledge of sheep and their breeders, and opening his address book to me. His help has been invaluable. I am also grateful to Phil Hadley, Senior Regional Manager of the Southern Region of Eblex, for his help with information on halal slaughter.

Of the farmers, shepherds and breeders, I thank: Alan Alderson, Nick Archdale, the late Robert Armstrong, Norman Bayley, Alec Bissett, Irving Blamire, Jim Brown, Susannah Coke, Michael and Julie Coney, Simon Dawson, Trevor Dodson, Jeff and Helen Dowey, Andrew Elliot, Mark Elliot, Tim Elliot, Huw Evans, Louise Fairburn, Francis Fooks, Rachel Godschalk, Stanley Jackson, Andrew Jones, Norman and Michelle Jones, Frank Langrish, Cyril Lewis, Brian Marrs, Frank Martin, Matt Mason, Angus Morris, June Morris, Philip Onions, Tom Patterson, John Ryrie, Bertie and Alice Thomson, Tim White, Lyn and Carol Williams and Lucinda Woodcock (who gave me the run of her extensive library).

I regret that I have been unable to include all the breeds and breeders I visited, but I hope they will be assured that nothing has been wasted, and everything I learned from the scores of people who helped me has contributed in one way or another to the making of this book.

Many friends encouraged me, some knowingly, others unwittingly simply by listening to me, particularly Rosemary Howell who has had to listen to me for more than thirty years. At one low point, Rose Peel cheered me up when she brightly said at least I wasn't in as bad a state as a friend of hers who had been 'broken' by writing a book. Emma Maxwell Macdonald regularly lifted my spirits with her enthusiasm whenever I fell into a slough of despond. Andrew and Carolyn Lewis encouraged and accommodated me in some style on my trips south. Miles and Glenis Postlethwaite freely donated their humour and trenchant advice, as did Peter and Laura de Wesselow. Pauline Dixon helped me to organise my disorganised thoughts. John Sadler advised me to 'just get the bugger written', and read parts of the manuscript even though he could hardly be less interested in sheep. Margaret Vane injected more enthusiasm into the project than sometimes even I could rouse, and Christopher Vane has, on many occasions, stiffened my resolve with words of quiet support. Alison Pilling administered doses of her life-enhancing zest and Sarah Edmondson her sobering realism. Nigel and Nina Lightfoot even tried to buy the book before it was written.

My children contributed in different ways: the idea first came to me while I was walking along the Northumbrian coast with Emily and my son-in-law Max; Libby helped me greatly, particularly with computer problems; Thomas read chunks of the manuscript and offered usefully honest criticism; and, as a result of his immersion in the ovine world, Edward now recognises, on sight, more breeds of sheep than most children his age.

It may be trite, but I mean it, that without my agent, Ed Wilson, and my editors at Profile Books, Rebecca Gray and Penny Daniel, and Mark Handsley, my copy-editor, this book would most likely never have seen the light of day.

There will doubtless be mistakes and infelicities; for all of these I take full responsibility.

Philip Walling
Belsay
January 2014

❧ INTRODUCTION ❧

... For there is good news yet to hear and fine things to be seen
Before we go to Paradise by way of Kensal Green.

G. K. Chesterton, 'The Rolling English Road'

THERE IS A PARALLEL WORLD AT WORK IN BRITAIN which most people, even those who live close to it, hardly ever notice and, even when they do, know little or nothing about. It's a world that has existed time out of mind and was once the foundation of all the wealth of England. And despite its decline from high importance in the Middle Ages and the ravages caused by the opening up of the New World, it continues, keeping faith with the passing seasons, obeying its own imperatives and adapting to survive.

This is the world of sheep husbandry.

Just as robins and blackbirds occupy the same territory and yet completely ignore one another, so most modern Britons occupy the same land as millions of sheep and, for all the notice we take of them, they might as well exist in another dimension. Everybody recognises sheep when they see them – they have woolly coats, live in fields, eat grass and have lambs in the spring – and there are few country places in Britain where you won't encounter sheep. But those things apart, to most people sheep are only sheep. They would not be able to name the breed, or the part of Britain it belonged to, or know what it was doing or why it was there. One of the purposes of this book is to try to remedy that. Another is to show you a little about something that we do really well here in Britain, something of which we ought justifiably to be proud. Because producing food and wool from our own soil is a real activity, not a metaphor, and unlike much that happens in modern Britain, it does not evaporate when you try to grasp it

How many of the millions of people who scurry past Shepherd's Bush every day give a thought to where the name came from? It was once an open heath, where sheep grazed under the eye of their shepherd, and there would have been a hawthorn bush trained into a shape that was once ubiquitous all across the downs and heaths of England. The thorn was pruned to grow into the shape of an oval cup, rather like an armchair; all the inner wood was removed and the outer branches were allowed to grow densely and knit together to a thickness of about eighteen inches. The trunk was shaped to make a kind of a platform upon which the shepherd could lay a bed of straw for comfort, and then he could step up onto it, throw a sack over the bushy sides to protect his arms from the thorns, and stand like a sea captain on his bridge, scanning the flock on the heath from his vantage point. Some shepherds' bushes were shaped to form a roof, as well as sides, and with a sack thrown over the top, would have made a fine shelter from the sun and rain.

Right in the middle of one of the most cosmopolitan cities on earth we have this permanent reminder of the enduring place that sheep have had in our lives. There are hundreds of sheep streets, sheep washes, sheep towns, associations with wool and weaving and spinning in every part of the kingdom. But these historical references and allusions should not lead us to believe that sheep belong to the past and have lost their importance. Far from it. We still have about 23 million of them in the United Kingdom, even though the national flock is reduced from 1992, when we had the ninth largest in the world, with 44.5 million head of sheep, half of them breeding ewes. It has never fallen below 20 million, apart from during and just after the two world wars when crop production took precedence. The sharpest reduction came in the year to June 2001, when the foot-and-mouth slaughter reduced the flock by 5.5 per cent. It has never recovered from this.

Britain and its people have been formed by waves of migrants flowing in over thousands of years. Historians can tell us something about their origins and the effect they had when they got here. But little is known about the domestic animals they brought with them. Perhaps they weren't considered worthy of recording, only being livestock, but the immigrants' animals, particularly their sheep, established themselves just as surely as did their human keepers. Some were crossed with the breeds they found here, others retained their purity for centuries, while a few have remained almost as they were when they arrived. By adapting to the landscape and climate, they found a place and established their character.

More than any other piece of land in the world, Britain is quintessential sheep country. Its climate and terrain are ideal for rearing sheep and sheep have been kept in large numbers throughout the British Isles for thousands of years. Our temperate climate, with little or no snow cover in most winters, and nearly always some vegetation available at any altitude, allows sheep to be kept outside in most places, throughout the year. From the thin soils and semi-tundra of the mountains of Scotland, Wales and the Lake District, to the Pennine fells, the rich lowlands of the Midlands, the marshes of Kent and the moors of the West Country, over many centuries, breeds of sheep have been developed which have become marvellously adapted to the land they live on. We have more than sixty different native breeds – a breed for every type of land and climate – and much of our landscape is the result of centuries of sheep grazing.

The French, who have remained closer to their soil than we have, talk much about *terroir* – the semi-mystical belief that the soil imparts a unique quality to everything that grows on it. Mention *le terroir* to a Frenchman and he will nod knowingly; no further explanation is required to explain, for example, the difference between wines made from the same variety of grape grown on adjoining plots of ground – sometimes only separated by a track or a stone wall – *mais c'est le terroir, c'est évident!* And our native sheep are just as much a product of their soil as are the Frenchman's grapes, or his cheeses or anything else that he gets from it. The soil imbues its products with certain characteristics. Sheep bred and reared for long enough on soils overlying limestone appear to take on a black or blueish bloom to their skin; whereas acidic soils, especially those containing iron, seem to impart a reddish hue. The best sheep-breeders are instinctively aware of these effects and strive, whether consciously or not, to enhance them. Breeders of the Swaledale aim for highly defined black and white skin and hair colouring; any shade of brown is not tolerated and they aim for a blueish cream fleece, like skimmed milk. By contrast, everything about the Herdwick, bred on the thin soils of the Lake District, tends towards steel-grey with reddish contrasting tones. And the Wiltshire Horn, on the calcareous soils of its native downs, is the quintessence of chalky whiteness.

But there is another thing about our national flock that is unique to Britain. We are the only sheep-keeping country in the world to have developed, over the last century and a half, a remarkably sophisticated stratified national meat-producing system, based on double cross-breeding, which has come to be called the *sheep pyramid*. The sheep at the top are pure-bred mountain and hill ewes, of which there are many millions, which form a genetic reservoir upon which the modern British sheep industry depends. These are moved downhill to better land and crossed with Longwool rams to produce breeding females which, in their turn, are crossed with Down breeds (referred to as terminal sires) to produce what are called butchers' lambs. The effect of this system is that most of the lamb we produce for the table in Britain is descended from one of our pure breeds of mountain or hill sheep.

However, we have not always had this pure-breed crossing system. Until about the beginning of the eighteenth century there were many different regional *types* of sheep, few of which were breeds that we would recognise today. All of the *types* were descended from four main wild prototype ancestors, and over long centuries they developed distinctly local characteristics. The *Urial* (or *Turbary*) ranges from the near east and eastwards into Tibet; its main distinguishing characteristics are a fawn coat, curved single horns in the rams and light erect horns in the ewes. They are prolific, twins being usual and triplets fairly common. Then there is the *Mouflon*, which was once widely distributed across Europe; the rams have massive horns that grow at right angles to the head, backwards (and sometimes outwards) and end in a tip just below the eyes after completing two thirds of a circle. The ewes are nearly always polled, i.e. hornless. The third is the *Argali* from central and northern Asia. Both sexes are horned but the rams' horns are larger than the ewes' and curl outwards with up to three spirals; they also have a distinctive white or grey muzzle (like the Swaledale). The fourth is the *Bighorn*, originating in north-east Asia and Siberia and later extending throughout North America, where, like their bovine counterpart the bison (and many other native species), they were hunted almost to extinction during the nineteenth century. Both sexes have horns, massive and curled on the rams. They have a hairy-woolly coat, like a Wiltshire Horn, a white rump like a roe deer and a white or grey muzzle like the Argali.

Their development into local types was partly through their adaptation to the soil and climate, emphasised by geographical isolation, and partly because breeders in a particular locality tended to favour a particular kind of sheep for sound practical reasons, such as there being a local market for its wool, or for its meat, or for its docility or fecundity, or whatever. Often a number of influences coincided and breeders enhanced certain characteristics because their experience told them that animals with those traits tended to thrive better than those lacking them and thus were more profitable.

But it is hard to trace the origins of our modern breeds because the evidence is lacking, incomplete or confusing or they have been difficult to classify. Some breeds manifest what could be described as primitive features – for example, six horns, or a carcase like a goat, or they automatically shed their fleece in spring – whilst others have been selectively bred for certain characteristics such as high milk yields or a particularly meaty carcase. There is also much scholarly disagreement over the routes by which different types of sheep came into Northern Europe. Advances in DNA analysis and carbon dating of bones have thrown some light into certain dark corners and added to the evidence, but there are still large gaps in our knowledge about the origins of particular breeds that will probably never be filled. There are many reasons for this: the literate ruling class (with notable monastic exceptions) tended not to concern itself with matters deemed proper for peasants. The origins and management of domestic livestock were not considered worthy of aristocratic concern. And the peasant flockmasters and breeders, assuming they were literate, tended to get on with their work rather than record what they were doing. Some of the later pioneers responsible for the breeding revolution in the eighteenth century were secretive about their methods and sometimes deliberately confused their rivals by misrepresentation. Breed names have been used imprecisely and interchangeably over the centuries and breed characteristics have changed radically according to demand, fashion and sometimes individual whim. And, rather like an elephant being hard to describe, even though we all know one when we see it, it is difficult to describe in words the differences between one breed of sheep and another.

Also, we tend to underestimate the effect of modern photography in reproducing a breed exactly as it looks. It was only fairly recently that livestock painters began to strive for realism: the sheep in William Taylor Longmire's 1870 painting of *Herdwick Sheep at Windermere, Seen from Low Wood* (kept at Townend in the Browne Collection by The National Trust) actually look like sheep, although not much like modern Herdwicks. And the pictures of the celebrated livestock painter Thomas Sidney Cooper, who knew livestock from a countryman's perspective, are accurate representations of animals as they were. But going further back in time, from the eighteenth century to the Middle Ages and beyond, when it would have been fascinating to see what domestic sheep actually looked like, such images as we have are far from realistic. Many of them were propaganda, done as caricatures to exaggerate desirable attributes, such as the 1842 picture of Jonas Webb's improved Southdowns, and an 1863 painting of Shropshire Downs, both of which depict sheep as preposterous blocks of meat standing on impossibly thin legs. There are some apparently accurate images from the second half of the eighteenth century, but they are remarkable for their realism.

We have also tended to take our sheep so much for granted that we forget that mankind has depended on them for much of our history. They are the essential domestic animal, more so than the cow, the goat or even the pig. They are also our oldest domestic animal and for centuries have satisfied many of our needs. Their tenfold purpose – meat, fat, blood, wool, milk, skin, gut, horn, bone and manure – provided us with food, clothing, housing, heating and light, all manner of domestic implements, soil fertility and parchment – which for centuries was the only material upon which a permanent written record could be preserved. Over the millennia each of these products has assumed a greater or lesser value as our needs have changed. We no longer use much tallow for candles, as we did during the eighteenth century, when the demand across Europe was such that the fat from a sheep's carcase was worth twice as much as the meat. Similarly during the wool boom of the Middle Ages the fleece was worth much more than the carcase. Now the carcase is worth between ten and twenty times the value of the fleece and the tallow is of negligible value. But throughout our association with them there has never been a time when we have not depended on sheep for one or other of their products. Our sheep represent a store of seasonal plant production that we can call upon when nothing else is available.

For centuries, when wool was our greatest cash crop, flockmasters kept sheep for the weight and quality of their wool. And the woolliest sheep were often the most ill-shaped, ungainly animals, slow to mature and living to great ages. Their breeding properties, carcase shape and fecundity were not all that important because only a few lambs were needed every year to maintain the flock size. Few animals were killed for meat and those that were tended to be older ones that had matured into the kind of mutton which would now be unattractive to modern palates. Many of these wool-bearing types were of venerable lineage, descended from sheep introduced into lowland Britain by the Romans to supply wool for their cloth manufactories. As the towns and cities grew, the demand for meat (and candle tallow) increased – although wool was still a worthwhile crop. Even into the 1980s the annual wool clip was reckoned to pay the farm rent. But as the urban population burgeoned in the nineteenth and twentieth centuries, for the first time in our long relationship with sheep we began to keep them almost exclusively to satisfy the demand for their meat. In the last decade or so, a fashionable niche market in sheep's milk products has opened up again in Britain for the first time in nearly a century. We abandoned sheep milking – mostly for cheese making – when liquid cows' milk became commonplace, unlike in Continental countries, where sheep's cheese continued to be made and sold in large quantities and cows' milk in bottles never caught on as it did in Britain.

This urban demand for meat caused a sea change in the British pastoral world over less than fifty years in the middle of the eighteenth century when a few farsighted farmers and graziers anticipated this revolution. The change in emphasis to meat marked the beginning of a long decline in the quality of fine English wool, and our renowned Longwools, such as the Lincoln and the Cotswold, and the incomparable Shortwools – notably the Ryeland – that had produced the wool-wealth of England in the Middle Ages were reduced to shadows of their medieval glory.

Then gradually, throughout the nineteenth century, the different regional types of sheep were developed into the kinds of distinct breeds we know today. Most of these were associated with a particular locality, but they were much more homogenously bred to conform to standards of breed uniformity than they had been when the emphasis was only on wool. By the second half of the nineteenth century the enhancing effect on their offspring of crossing together certain pure breeds became more widely recognised (particularly after the work of the monk Gregor Mendel in cross-breeding peas). Then when the railways made it easy to move livestock over long distances at a fraction of the cost of droving, the way was open to British sheep farmers to develop the sheep pyramid, the sophisticated national meat-producing system that we have today.

This book is an attempt to give a flavour of the wonderful

story of how we and our versatile, compliant companions made our landscape in the great endeavour of taming the wilderness. For man and his sheep stand in partnership outside wild nature, on the side of the civilised world, transforming its vegetation for human benefit. Perhaps the most exciting thing is that the whole pastoral history of our Islands can be traced through breeds that still graze our pastures. They are all still here. So let's start at the beginning.

THE SHEEP OUR ANCIENT ANCESTORS KEPT

... *Theirs is no earthly breed*
Who only haunt the verges of the earth
And only on the sea's salt herbage feed –
Surely the great white breakers gave them birth ...

Roy Campbell, 'Horses on the Camargue',
from *Adamastor*, Faber & Faber, 1930, p. 80

A BOUT THIRTY YEARS AGO MY NEIGHBOUR ACQUIRED a dozen piebald Jacobs (gentlemen's parkland sheep, as I thought) with two, four and sometimes six horns, to graze the fields around his house. He was a lecturer in engineering and had designed his own elaborate sheep pens to hold his little flock for handling. He had also acquired a collie dog. But this dog flatly refused to have anything to do with sheep. If she even got wind that she was expected to go anywhere near them she would run into her bed and refuse to come out. No amount of cajoling, or even dragging, made the slightest difference to the recalcitrant animal.

From time to time his wife and daughter and anyone else who happened to be at the house were enlisted to try to pen up his Jacobs, but for eighteen months he failed to get them to go into his pens. They simply refused to be driven anywhere near them and whenever he tried to corner them they scattered. It was like a circus. They all gave birth during their first lambing time without needing any help, so he left them to fend for themselves. They were not shorn that year and by the next spring they were a sorry sight, carrying two fleeces, one on top of the other.

However, there were compulsory dipping regulations in force. And after their second lambing, when the flock had grown to over thirty mixed females and uncastrated males (which formed a separate little pack away from the ewes), he had to find a way to get them into his pens to comply with the law. So he asked me if I would go round with my dogs and help him.

Now these sheep were wild. I only had to rattle the chain on the gate and their heads went up. As the dogs circled round them they split up into two groups: one flocked on a little hillock at the top of the field and the other made for a gap in the hedge into the next field. I put one of my dogs round the escapees and with some difficulty she brought them back through the hedge into the field, but as soon as they realised they were being manoeuvred towards the sheep pens, they divided themselves into another two troupes, one which allowed itself to be eased into the pens, but the other legged it for the opposite corner of the field from the little group on the hillock. The more we tried the more agitated they got.

This charade went on for ages, with me getting more and more irritated and the dogs more and more desperate. These sheep had never been bossed or dogged and with each failed attempt to get them into the pen they grew bolder, leaping over the dogs and running away like deer with their leggy little lambs keeping close to their mothers. My neighbour's daughter and wife emerged from the house roused by our shouting. The four of us, with my two dogs, eventually got the flock penned, but one ewe and her lamb leapt the wooden railings and took off down the road towards the village. We later cornered them both in a garden a mile and a half away. They were completely exhausted or we wouldn't have been able to catch them, but even so, the ewe still had enough fight left in her to jump about as I dragged her by a horn towards the van, with her frightened lamb tagging on behind. But as soon as I got her into the van the lamb turned tail and took off down the road as fast as it could run. I sent the dogs, but they couldn't turn the terrified lamb, and after a hell of a chase, it finally gave up and flopped down on the verge, panting and lay there. I picked it up and threw it in the van with its mother, who by this time had set up a tremendous bleating and was running at the back windows butting them, trying to break out. This was my first encounter with primitive sheep and it rather prejudiced me against them.

Apart from oddities like Jacobs, there are two main primitive types of sheep, which came into Britain by two different routes. One is the northern short-tailed group (which for shorthand I call 'Viking' sheep). These came into northern Britain via Scandinavia and Russia from central Asia. By a genetic quirk, they only have thirteen vertebrae in their tails, compared with twenty in other sheep. The second type is a long-tailed Celtic sheep believed to have come from the Near East through the Mediterranean into southern Britain and then spread north. An example of this Celtic type is the Soay that roams semi-wild on Hirta, the largest of the abandoned St Kilda islands. But its near neighbour, the Hebridean, is a short-tailed Viking sheep. It seems that the two early migrations reached the furthest extent of their ranges in the Western Isles, where they met but never mingled. Rather as the Isle of Barra is Catholic and next door the Isle of Lewis is fiercely Protestant.

One of the most unusual of the Viking sheep is found on North Ronaldsay, or Rinansey (Ringan's Isle) in Old Norse, the most northerly of the Orkney Islands, their last redoubt in Britain. They have endured here, on the very edge of the British Isles, because their island is so isolated and for the last two centuries have been confined to the foreshore for a large part of the year. Here they fend for themselves, and have adapted themselves to living on a diet of seaweed.

They do most things the opposite way round to other sheep. At their best during winter, when the red seaweed they prefer, *Palmaria palmata*, or 'dulse', is most abundant; they can't eat grass for too long, or they are poisoned by the copper in it, yet they cannot get all the annual sustenance they need from seaweed alone; and unlike other sheep, which eat by day and chew their cud at night, they feed according to the tides, lying up on the foreshore at high water and then following the ebbing tide onto the rocks to graze the exposed seaweed. They can even swim. Some of the most intrepid will plunge into the ebbing seawater and head for an outcrop to be the first to reach the tastiest fronds. They are as agile as goats, negotiating the slippery rocks, unafraid of the surging tide. And they are not entirely vegetarian. They have developed an odd partiality to the feet and legs of dead seabirds. When the new automatic revolving lighthouse at Dennis Ness was installed, it attracted flocks of birds, which flew into it and were killed. Their carcases proved irresistible to the sheep, which came from all around the shore to eat their legs.

Although the sea sustains them, it is an exacting benefactor. For in winter, powerful Atlantic tides surge around the island, in contention with the calmer waters of the North Sea. And when a westerly gale blows against a running tide over the shallow uneven seabed, many of the smaller low-lying islands, such as North Ronaldsay, are often ringed by broken water and overblown by spindrift for days on end, confusing land and sea. During particularly violent storms, the little sheep have hardly any protection from the crashing waves that douse them with salt-water and spray, and even on calm days the land is seldom free of a ruffling breeze.

The island's flock was banished to the foreshore in the 1830s because the islanders were in desperate circumstances. Seaweed grows in vast quantities in the cool coastal waters around North Ronaldsay, nourished by the Gulf Stream. The crofters had used its almost unlimited growth as fertiliser for their sandy soil, dried it for fuel (there is no peat or wood on the island) and in winter, when other fodder was scarce, supplemented their animals' diet with it. But there was still a vast annual crop that went unused.

So when James Traill, an Edinburgh lawyer, bought the island in 1727 for 2,000 guineas (£1 an acre) he probably had more than half an eye on the huge potential for kelp making. His purchase included the foreshore, which gave him the right to gather the 'tangles', as the seaweed *Laminaria digitata* is called locally, to make the kelp. Traill made it a condition of the crofters' tenancies that they collectively produce a certain tonnage of kelp each year. It was an arduous business. Between forty and fifty cartloads of wet tangles made a ton of kelp after being dried and burnt, in controlled fires, like charcoal-burning, in shallow pits on the shore. A ton of kelp would make about 8 lb of iodine.

During the fifty years between 1740 and 1790, kelp brought in about £37,000 to North Ronaldsay. The crofters produced about 150 tons a year, at an average price of £5 a ton, which rose to £20 during the Napoleonic wars. Half the income went to the crofters, the laird received a third and the balance went on shipping and expenses. But this large cash income had a corrosive effect on the crofters' moral economy, inducing them to neglect their crofts and fishing and live entirely on the profit from the kelp trade. The prosperity which it brought to the island caused its population to grow in fifty years by 40 per cent from 384 to 522.

In 1793 Sir John Sinclair of Ulbster in Caithness (first President of the Board of Agriculture and the man responsible for bringing Cheviot sheep to the north of Scotland) warned of the foolishness of relying on the kelp trade, 'agriculture, which in every county is the first and foremost of the arts, is greatly neglected; and a style of living has been introduced among the proprietors, which their lands can by no means support, and which, if ever this manufacture should fail, must bring certain ruin upon them, their tenants, and their families'.

Sinclair was proved right: by 1832 the price of kelp had collapsed, and, with the islanders facing destitution, even famine, something drastic had to be done. Traill's grandson, the then laird, drawing on his experience of colonial administration in India, together with his land-grieve (agent), Robert Scarth of Sanday, known as 'a mesterfu' man', came up with a radical plan to save the island's people from extinction.

Apart from encouraging emigration to larger, less populous islands, the plan involved 'land-squaring' and building a dyke round the island to keep the sheep off the cultivable land across which they had hitherto roamed unchecked. Up to then the whole island was cultivated in a Scottish version of the communal open-field system (the 'run-rig') that had been abolished in much of the rest of Britain by this time. The crofters were allotted strips of the island's arable land each year so that everybody got a share of the best and the worst. Land-squaring abolished the run-rig and divided the land between the crofters to make it more productive and banishing the sheep prevented them from roaming over the crops and damaging them. The island's flock was still kept mainly for its wool (and a little tallow for lights) and only occasionally killed for meat, because as is usual in pastoral societies they seldom ate their sheep (and when they did it was only the older ones).

The drystone sheep dyke was twelve miles long, built round the island above high water, to 'louping height' – about six feet – a little higher than the sheep could jump. It separated the 270 acres of foreshore from the rest of the island. The crofters were then allotted the right to keep a certain number of sheep in the communal flock, according to the size of their squared-up holdings. The flock has been confined to the beachhead and foreshore ever since, although the ewes are allowed into the fields for about four months between lambing in mid-April and weaning in late July or early August. The nutritive value of seaweed varies with the seasons, but its great benefit is that it is at its best in mid-winter, when it is most plentiful and there is little else to eat. In spring, at lambing time, it also has milk-stimulating qualities.

The paradox is that Scarth's scheme was not intended to benefit the flock, which was considered something of a nuisance, and of marginal value to the island's economy. Had the flock not been banished to the foreshore it is unlikely it would have survived because the crofters would probably have sacrificed the sheep to grow the crops which they desperately needed to sustain the population. Although it was not a new experience for them to eat seaweed, forcing the flock to subsist on it most of the year was a unique experiment, and not certain to succeed. It is probable that there were many sheep unable to adapt to the new regime which simply died, and the current flock is descended from the survivors.

Scarth established the Sheep Court to regulate the management of the flock according to rules agreed between the laird and the crofters. Each of the five 'toonships' elects two Sheepmen to the Court to enforce the Regulations, and the Court appoints a secretary. In Scarth's time there were seventy-one crofts, keeping 2,250 sheep. Now the flock is between 2,500 and 3,000, because there is an abundance of seaweed, which is no longer used for anything other than feeding sheep, but there are far fewer crofts.

As with all the short-tailed types, these little sheep are

very prolific: three-quarters have twins, triplets are common and even quads not uncommon. The lambs are nearly all born within four weeks in April and May, and are very small at birth, little bigger than kittens, so there are hardly ever any lambing problems. But despite their prolificacy, it is traditional on North Ronaldsay to allow ewes to rear only one lamb each. The rest are killed shortly after birth. They prefer to kill the gimmer (female) lambs unless they are wanted for replacements, because the wethers (castrated males) make a bigger carcase – about 15–20 kg deadweight. So for every 100 ewes, they keep about fifteen gimmer lambs and kill all the rest.

This sinister business is never mentioned in any of the books or travel articles. The National Sheep Association handbook on *British Sheep* is silent on the subject, as was nearly everybody I spoke to. Even nineteenth-century writers do not refer to it. It is possible that the killing was not as extensive as it is now, although I doubt it. Shepherds usually go round the lambing fields trying to save lambs' lives. Here they go round with an iron bar. Far more lambs are killed than reared because it is commonly believed that each ewe is only capable of rearing a single lamb. This is curious because the North Ronaldsay's short-tailed cousin, the Shetland, can rear multiple lambs. It might simply be the crofters' prejudice, or the sheep's diet, or poor milk supply, but it can't have anything to do with the size of the island because even if they kept half the number of sheep, the crofters would still believe each ewe only capable of rearing one lamb.

As the propensity for multiple births is inherited and most of the island's sheep are twins and triplets, these will tend to breed twins and triplets themselves. If the crofters were to breed only from ewe lambs that were singles it might be possible to breed out the fecundity, but that would require either a communal effort recording and marking the sheep, or individual owners managing their own sheep separately from the rest, which is impossible under the present system.

Individual owners cannot even select rams for breeding with their own ewes, because all the sheep run together on the shore. The only control over breeding is to castrate all the ram lambs except the few kept for breeding. The rams left entire and running with the flock all year have a very narrow window of breeding opportunity, because the ewes only come into season for a few weeks each November and December; otherwise they are unreceptive to the males, which tend to keep away from them in a separate group.

One result of this breeding free-for-all is that the sheep carry a remarkable range of wool colour, from the chocolate tones of moorit, to steel-grey, black, cream and white. Every animal is an individual, and for modern commercial farmers, who strive for standardisation, this would be anathema. But the crofters delight in their sheep's individuality and do not want animals that all look the same. They have a fine inner fleece for warmth, and an outer protective layer of wool to keep out the weather. Ancient Iron Age sheep would have had a similarly wide range of fleece colour and would have looked like these sheep before weaving began to demand white wool that would hold a dye.

Shearing is done with hand shears that leave an inch or so of new wool, impregnated with lanolin, which repels the weather and protects the sheep. The machine hasn't caught on because it leaves the skin too bare for them to withstand a salt-water drenching soon after clipping. Shearing is done at the first new moon at the end of July or beginning of August. It was once common practice in agricultural and pastoral societies to work with the movement of the planets and wait for the most propitious celestial time to carry out important annual tasks. In the West we have largely abandoned what we have come to believe is superstition. But not on North Ronaldsay.

They only recently gave up the ancient practice of removing the fleece by *rooing*, which involved pulling off the sheep's
wool by hand. Even in the past many thought it a cruel practice, but now it seems positively barbaric. Professor Low of
Edinburgh University was not impressed when he wrote 200
years ago in *Fauna Orcadensis* (1813, p. 7) that 'about midsummer all the men in the parish attended with their dogs and
gathered up the whole flock ... into narrow pens, and from
thence I may say to the place of execution, where the wool
is torn off their backs; an operation which brings their whole
blood into their skin, and is not only disgusting, but, if the
season proves harsh, is the cause of great destruction. But
however cruel it may seem, it is almost the only notice that is
taken of these useful animals by their unfeeling masters until
that time twelvemonth.' But it may not be as cruel as it seems,
because there is a natural annual break in growth between the
old wool and the new, and at the right time the old fleece will
come away fairly easily from the inch or two of new growth
underneath. In fact it is the only way of removing all last year's
growth without taking any of the new wool. I imagine it must
be less painful than waxing, or even ripping off an Elastoplast.

When on the shore the sheep naturally divide themselves
into 'clowgangs', separate family groups that have claimed a
piece of the beach-head and foreshore as their territory and
which they defend from interlopers by 'doosing' (butting)
them off. They are perpetually restless, like the sea, moving
endlessly up and down the shore, combing the rocks and
shore; seaweed will not sustain them entirely and an exclusive diet of grass poisons them. The wethers in particular are
affected by ingesting the minerals in seaweed, which causes
stones to block their urinary tract and nearly all the wethers on
the shore seem to dribble urine continuously, like an old man
with prostate trouble. They instinctively know the state of the
tide and, as if summoned by an alarm, within a few minutes

of it turning they rise to their feet, stretch and begin to follow the retreating sea down the rocks, competing with one another to be the first to reach the crisp clean blades of ware as they are exposed.

Six times a year the flock is *punded*. This is a communal gathering of the clowgangs into one of the nine punds, or sheepfolds (English 'pound', as in police pound), built at intervals round the island beside the wall. The pundings are done at high water while the sheep are confined to the beachhead. The two at Christmas Eve and New Year's Eve are to draw out the 200 or so three- or four-year-old wether sheep for sale for slaughter. In the days when they didn't sell so many wethers, the crofters used to get their dogs to catch each one on the shore, by grabbing the wool at the back of the neck and holding on. The February punding is to 'score', or count, the sheep to make sure nobody has exceeded his quota. There is one at midsummer and another a fortnight later for rooing – now shearing. The last one of the year is for dipping in the autumn.

The wethers are not ready to slaughter until they are three years old. By which time their meat is really mutton, dark, pungent and too sheepy for many modern palates. Some attempt has been made to sell it as a gastronomic delicacy through two outlets, Orkney Meats and a butcher in Kirkwall, with the evocative name of Thorfin Craigie. And they have had some help from Mey Selections, Prince Charles's marketing company run from the Castle of Mey, across the Pentland Firth. But 200 carcases a year is hardly enough to create a market. If this were France the meat would be given an *appellation contrôlée* and sold as a gourmet product, available for only a short time in the winter, like Vacherin Mont-d'Or.

To find a living example of the Celtic type that came into Britain by the southern route, we have to go to another island, at the opposite end of the country. Here a breed whose lineage reaches back into the Iron Age survived into the last century. Its forebears once grazed across the heaths and downs of south-west England in a great family of tan-faced, horned sheep, whose modern descendants are the Dorset and Wiltshire Horn. Long after this type had disappeared from the mainland it survived on the Isle of Portland, living on sparse pasture, and was renowned for its sweet, delicate mutton, rich cheese and fine wool.

Portland has been a royal manor since before the Conquest and a considerable sheep-run for much longer, famous throughout the kingdom for its unique breed. Although it has not been an island proper for 10,000 years or more, it is only tied to the Dorset mainland by the shingle bank of Chesil Beach, and has been enough of an island to keep its sheep from any surrounding influence. This ensured that the Portland breed remained pure through many centuries, during which those who could afford it esteemed its flesh 'as fine flavoured as any in the kingdom' and paid nearly twice the price that they paid for mutton from other breeds. For many years the Queen's Own Yeomanry held its annual summer camp at the Gloucester Hotel in Weymouth just so that the officers could dine on Portland lamb and Portland new potatoes.

The conventional view is that this tan-faced type descends from similar Northern European stock to the Orkney and the Hebridean, but I am not so sure. They have a number of traits that are hard to square with their being of northern provenance. The ewes possess the unusual capacity (inherited by the Dorset and Wiltshire Horn) to lamb out of season, when the northern types only take the ram in the autumn and lamb in the spring. All-year-round breeding is an inherited trait of Mediterranean sheep, as is the almost complete absence of multiple births. This was noted by writers in the eighteenth century and has not changed, even when the ewes are moved to better going just before conception – a process known as 'flushing'.

As fecundity is inherited, pure-bred sheep from single lambs tend only to produce single lambs. The fact is that if ewes are kept in hard conditions they usually only have single lambs because when they ovulate either only one egg is presented for fertilisation or more than one egg is either reabsorbed or dies during gestation. This is considered to be an example of natural selection because in hard conditions single lambs have a better chance of survival. Then, over generations of breeding from single lambs, the tendency to have only one lamb becomes fixed in the breed. But the northern short-tails have endured almost perpetual hardship and they still produce an average of two and a half lambs per ewe. The Portland's poor lambing rate is slightly mitigated by its capacity to have three lambs in two years, but this is not enough to make it fit into the northern type – not least because it is not short-tailed.

Folk memory says Portlands swam ashore from a shipwreck – possibly the Armada. But then so many of our sheep breeds are credited with a similar origin – the Herdwick and the Jacob being examples – that, as was noted in an article in *Country Life* in May 1953, 'one begins to wonder if Noah had not passed by [the coast of Britain] in his Ark'.

Sir Henry Harpur, sixth baronet, visited one Farmer Lowman in 1770, to buy a troupe of Portlands to graze his newly extended park at Calke Abbey in Derbyshire. He had tried the New Leicesters because they were the height of fashion at the time, and their creator, Robert Bakewell's farm at Dishley, was only twelve miles away, but the Leicesters were high-maintenance, demanding sheep, used to better feeding from the heavy clays of the Trent and Soar valleys, and not at all what was needed on the thin calcareous soils at Calke.

Doubts were expressed at the time about the wisdom of

removing sheep to Derbyshire from their coastal territory. It was considered that flocks naturally adapted to their terrain would not easily thrive away from it. This is probably right generally, but the fundamental nature of the land plays a part, and the underlying limestone at Calke is not dissimilar to that of Portland, even though the winters are colder away from the softening influence of the sea. Anyway, Sir Henry's choice was vindicated because the Portlands adapted so well to their new home in Derbyshire that his grandson, Sir George Harpur Crewe, returned to the Isle in 1835 to buy some more from Farmer Lowman's grandson. He records in his journal that when he arrived, there had been no rain for six weeks, there was hardly a blade of grass to be seen and the Isle's 3,000 sheep appeared to be living on thistles, 'of which there was as fine a crop as ever I saw'.

In her 1911 book *Shepherds of Britain*, Adelaide Gossett remarks on the Portland's remarkable resilience, despite 'being roughly attended to'. They roamed the Isle by day, with the leading animals wearing bells to keep the flock together and so the shepherds could find them amongst the cliffs and rocks in the evening. And although primitive breeds are distinguished by their strong resistance to disease and parasites, the Portlands were in a class of their own. This strength of constitution tends to diminish as sheep become *improved* and more productive.

Miss Gossett tells us they were not afraid of sheep dogs; they resented the collie and would butt him, so the shepherds used what she calls the 'English sheepdog' (presumably the rougher hairier type of dog from which the Old English has been bred). Like many of the other ancient breeds, these are proud sheep, independent and not easily intimidated, with a sense of their own dignity, and a bearing that demands respect. They are not the slaves of commercial farming, to be used and abused to produce meat, but come from an earlier time when relations between people and their animals were regulated by a greater respect for the servitude that domestic animals give their human keepers.

Portlands eventually disappeared from their island because it was gradually being eroded by the commercial quarrying of its eponymous and valuable stone. Sir Christopher Wren was MP for nearby Weymouth when he caused over 6 million tons of Portland stone to be used in the rebuilding of St Paul's Cathedral after the Great Fire. The eastern façade of Buckingham Palace, innumerable public buildings across London, and the Cenotaph are all made from the stone, as are the headstones and memorials to two world wars.

The last 'scraggy miserable lot' from the Isle was sold at Dorchester market in 1913 and even the persuasion of the auctioneer, says the newspaper report, 'did not avail much to make the final bid anything but derisive'. By 1953 only a few small flocks remained, scattered around the country and mostly kept by eccentrics. Few commercial farmers would countenance a breed, no matter how sweet and delicate their flesh, that barely averaged more than one lamb from each birth and whose carcase, even after two years, hardly reached 50 lb dead weight. Calke Abbey was the largest remaining flock, but not considered pure because over the years it had received infusions of blood from other breeds, notably the Exmoor Horn (a near relative).

Across the bay from Portland there is a little flock of twenty ewes and followers kept by Norman and Michelle Jones at Hogchester Farm near Lyme Regis. I was struck by how small they were, and square, with wide-set legs and impressive horns, heavily spiralled in the rams and curled into a delicate half-circle in the ewes, with a characteristic black line running through the leading edge of each horn. They were not in the least timid, and I thought I saw a flash of ancient pride in the way the ewes stood their ground, and in the way the rams will run at you and try to knock you down if you get too close. Their remarkable little lambs are born foxy-ginger and, as they mature, gradually grow a creamy-white fleece of very fine wool. But they never lose a gingery tannish hue on the hair of their faces and legs.

There is something wistful about the modern interest in rare breeds of livestock. None of the commercial breeders in the past cared a jot for retaining breeds that had outlived their usefulness; nor do they much care now – that's largely why there are hardly any accurate records of what the early types looked like or how they were transformed into modern breeds. Sheep breeders produced what would satisfy demand and were not concerned to keep types that had outlived their usefulness, unless nothing else could survive on their land. And that is the reason why the Portlands held out on their rock where few other breeds could have lived. The thin soils, the salty influence of the sea on their sparse pastures, and a system of common grazing that had died out much earlier on the mainland would have done for most modern breeds.

Most sheep are either polled or grow two horns; in many breeds, the males are horned and the females polled. But there are primitive breeds that grow four, six, even eight – and occasionally odd numbers of horns, three or five. We see this in the Hebridean, which has only just survived through to the modern age on the edge of the British Isles. It is part of the same northern short-tailed tribe as the North Ronaldsay and the Orkney, and was once found all along the western seaboard of Scotland from the Hebrides, into south-west Scotland, down to the Isle of Man and into the Channel Islands. They are double-fleeced like all the northern short-tails, with a fine inner coat and a harsher weather-proof outer coat, usually of black or dark brown, becoming grey (particularly round the muzzle) as they age. Multiple-horned rams are hard to keep alive because when they fight (and these rams find it irresistible) they easily split open their skulls. For this reason it is common practice to castrate males in park flocks.

They are now ornamental sheep, because apart from the superior quality (but small quantity) of their wool, modern breeds grow quicker, rear heavier lambs and respond to better feeding and more intensive management. The great value of all the primitive breeds is that they are superbly self-sufficient, hardly ever having trouble lambing, and are very hardy, with natural resistance to parasites and diseases. And have a remarkable capacity, which most improved breeds have lost, of extracting from the coarsest grasses and vegetation whatever goodness they contain. They actually seem to prefer grazing purple moor grass (*Molinia caerulea*), which other breeds won't touch, and birch scrub, in preference to heather.

Because of this proclivity they are valuable companions in moorland regeneration projects where coarse grasses have to be controlled to allow heather to regenerate. The point here is that these older breeds, better adapted through long survival to live on poor vegetation, are often the only way mankind can obtain anything of value from otherwise unproductive land. They are great survivors in places where other grazing animals would not last a week. If it is possible in the pastoral world to get something for nothing, breeds like this offer the best chance of doing so.

They can live all year round on what vegetation they can pull, and even scratch through a foot of snow to get to it. They are long-lived, often producing ten or more crops of lambs compared with half that number in the improved breeds. And their flesh tastes rather like game, well-flavoured, lower in saturated fat than the modern meat breeds, and although the carcase is small compared with modern sheep, smaller portions satisfy the appetite.

One of these types is the Manx national sheep, the Loghtan, *lugh dhoan*, which in Manx Gaelic means 'mouse-brown'. Its wool is actually moorit (chocolate brown) and not much like the colour of any mouse I've ever seen. Extravagantly horned and slow-growing, these sheep had been naturalised on the Isle of Man for many centuries, but by the 1950s they had dwindled to only about 100 animals. Unlike smaller islands where breeds survived because they were by-passed by the agricultural revolution, the Isle of Man is big enough for its farmers to run serious commercial enterprises and the backward local sheep did not fit in with this. No farmer who wanted to be taken seriously would have considered keeping the Manx, or any other primitive breed.

But all things come round again if they are not destroyed in the meantime. So when in the late 1980s George Steriopulos saw these animals in the auction mart he wondered why nobody wanted them. He bought some, and with his wife, Diana, has followed in the footsteps of Jack Quine, a Manxman, who almost singlehandedly kept the breed going through the lean 1960s and 1970s. The Steriopuloses have spent three decades farming the Loghtan back from the edge of extinction on the Isle of Man, building up to a flock of about 1,000.

During the 2001 foot-and-mouth epidemic they were so concerned that the breed might be wiped out that they kept part of their flock in isolation on the Calf of Man. Then in 2008 they obtained European Union Protected Designated Origin status (*appellation contrôlée*), which puts the breed in the same category as Parma ham, Stilton cheese and Champagne. Progressive farmers are wrong to be contemptuous of their local sheep. They are not only a Manx symbol, depicted on their coins, but their unique characteristics make them a paying proposition on poor land, giving them a value over and above the blandly conformist modern breeds that everyone else keeps.

Some breeders have been so enamoured of the qualities of the ancient breeds that for aesthetic reasons, combined with a veneer of practical justification, they have created their own vanity version. In the early 1900s Sir Jock Buchanan-Jardine

bred the Castlemilk Moorit from a cross between a Shetland,
a Manx Loghtan and the wild Mouflon. He wanted it to grace
his parkland in Dumfriesshire and also to provide woollen
clothing for his estate workers. The Mouflon gives it its distinctive white markings on its underbelly, rump, lower jaw, knees
and inside lower legs and round the eyes. After Sir Jock died the
flock was only saved from extinction by the redoubtable Joe
Henson, founder member of the Rare Breeds Survival Trust,
who at the dispersal sale in 1970 bought a ram and nine ewes.
From those ewes the current few hundred have been bred and
this unique modern manifestation of an ancient breed was
saved. The moorit colour is recessive, so the first cross with any
other breed nearly always produces wool of the same colour
as the dominant crossing breed. It is also naturally short-tailed
with long legs which make it a very elegant mover across the
ground.

But the enigmatic Jacob is different from the other primitive breeds and its origins remain tantalisingly obscure. Its
fine bones, slender body and lean carcase, even its tendency
to grow multiple horns, set it firmly within the primitive
camp. The gene for multiple horns is linked to a condition
called split-eyelid. In its mild form the upper eyelid has a
v-shaped nick in it, and at its worst the eyelid is divided into
two, causing considerable irritation and distress. But the Jacob
has remained aloof from the evolutionary links and changes
that have affected other primitive breeds and has kept itself
as a pure, direct link with the ancient world, unchanged since
its creation, some would say since The Creation. As with the
world's other domestic sheep, its most likely birthplace is the
Levant, where a piebald sheep existed more than 3,000 years
ago, known as Jacob's sheep and whose origin is explained in
the Book of Genesis (30: 31–43).

Jacob had toiled, without wages, for his uncle (and father-
in-law) Laban, for fourteen long years, for love for his wife

Rachel ('ewe'). He agreed to continue in Laban's service on condition that he would be allowed to keep as his share of Laban's flock every 'speckled and spotted' sheep. These are described as the 'sportings of nature', unusual and spontaneous variations from its regular production. There cannot have been many of these as otherwise the acquisitive Laban would not have agreed. Jacob took these sheep 'three days' journey distant', to keep them separate from Laban's flock, and continued to shepherd the rest of Laban's sheep, which were of a uniform colour – probably brown or black.

Then Jacob did something that in ancient times accorded with the commonly held mystical belief that whatever the female is looking upon at the moment of conception will affect the nature of her offspring. He took some fresh saplings and 'pilled white strakes in them' – exposed in streaks the white pith beneath the bark. Then he set these sticks up at the watering troughs where Laban's sheep mated when they came to drink. From these matings a few of the ewes produced speckled and ringstraked (streaked in rings) lambs. Then at their next conception Jacob 'set the ewes' faces towards the ringstraked' so that they might produce more of the same. Jacob only chose the stronger animals to breed his speckled flock from, so that gradually his speckled and spotted sheep became dominant in the flock, leaving Laban with the weaker pure black or brown ones.

Laban was none too pleased by Jacob's crafty behaviour and he tried to limit the kind of sheep that Jacob could have, first to speckled, but when the coloured lambs that were born next year were all speckled, to ringstraked lambs, but of course the next crop of coloured lambs turned out to be ringstraked. And so, over time, Jacob acquired all Laban's sheep as they gradually became multi-coloured. Then by selecting those sheep with more white in their fleeces, Jacob's sheep gradually became white. So that by the time of King David (as told in

Psalm 147) their fleeces are compared to snow; and in the Song of Solomon he sings of his mistress's teeth being like a flock of sheep just come up from the washing.

But the Jacob sheep that has come down to us is still spotted and piebald, whereas through divine favour Jacob's flocks lost their multi-coloured fleeces and became pure white. The breed was not called the Jacob in England until the twentieth century, before which it was known as the 'Spanish sheep'. One long-established flock in Warwickshire, the Charlecote, claims a Portuguese origin, but it is unclear whether the evidence of importation (in a letter from 1756) refers to the then existing flock or to an addition to it. There are numerous eighteenth-century references to four-horned piebald sheep grazing gentlemen's parks: Robert Bakewell was well aware of them because he wrote to Arthur Young in 1791 asking if he could procure some of 'the four-horned kind ... called Spanish ...'

Jacobs are yet another breed supposed to have swum ashore from a Spanish galleon in 1588. But the English used the word Spanish to describe anything foreign, exotic or dubious: Spanish influenza, Spanish practices, Spanish juice (liquorice) Spanish fly (the aphrodisiac), and so on. Spanish is used to indicate foreignness, often in a grudgingly admiring way. These were gentlemen's sheep, and commercial farmers would be disdainful of their being kept as ornaments with no concern for profit. To their gentle owners they were living lawn mowers that bred their own replacements and needed no fuel. But to a working farmer they were (and still are) little better than goats, and 'a damned nuisance'. It is most remarkable that they have resisted 'improvement' by crossing with any of the breeds that have come in and out of vogue over the last two or three centuries. There is something rather magnificent about this, which I was too young to appreciate all those years ago, when I lost my temper with my neighbour's silly sheep.

℘ 2 ℘

THE ROMANS AND
THE WOOL BREEDS

If wool growing is your business, beware of barbed vegetation,
goose grass and star-thistle: avoid too rich grazing:
choose from the start, a flock both white and soft of fleece.
Reject any ram, however pure and white his wool,
if the tongue beneath his moist palate is black, for he'll breed
lambs with black-spotted fleeces – Reject, and look around for
another ram on the crowded sheep run.
With the lure of such snowy wool, Pan, god of Arcady,
tricked the Moon …

Virgil, *Georgics* (3.355)

L OUISE FAIRBURN WAS MARRIED IN A WEDDING DRESS
made from the fleece of one of her Lincoln Longwool
sheep, the men wore waistcoats of the same wool
and they dined on meat from one of her lambs. She invited
the newspapers to write articles about her eye-catching
wedding and they took photographs; in one, Louise is standing with her prizewinning ewe, Risby Olivia, high up on her
land on the edge of the Lincolnshire Wolds; at her back the
glorious panorama of the Trent valley stretches away to the

west across the sweep of fertile fields where the multitudes of woolly sheep once grazed that made England rich.

One old Lincoln breeder with a Lincolnshire pedigree as long as his flock's, with thirty generations of breeding behind it, was rather critical of people who fall in love with these admittedly endearing, woolly sheep, and are newcomers to sheep keeping who don't really know how to manage them when things get difficult. He rather disparagingly referred to them as 'ladies with a paddock'. So far Louise seems to be proving him wrong. And her masterly publicity coup for her little flock of Lincolns, and for the Lincoln breed in general, has done much to bring the Lincoln to public notice.

The Lincolnshire Wolds are so deserted and quiet now it is hard to imagine the wealth that once flowed into this county from all over Europe. Beginning with the Romans, and then growing to a flood throughout the Middle Ages, Lincolnshire produced some of the best wool in Europe, which made it one of the richest counties in England and paid for some of the finest medieval churches ever built in Christendom.

There are ten English Longwool breeds, now much diminished in importance from the glory days when their wool was the mainstay of the English Exchequer. The Leicester Longwool was used by Robert Bakewell in the eighteenth century, to create a breed that bridged the change between medieval wool and modern meat production, and afterwards fell into oblivion. Four of its descendants, the Border Leicester, Wensleydale, Teeswater and Bluefaced Leicester, have matured into important crossing breeds. Two are moorland Longwools, the Whitefaced Dartmoor, which is the only Longwool breed in which the rams are horned, and its cousin the Dartmoor. Their neighbour the Devon and Cornwall Longwool produces the greatest amount of wool from each sheep of any Longwool. The large Cotswold and its close relative the Lincoln are almost indistinguishable and have sunk into obscurity with no modern role other than perhaps to show us what a real Long-wool looks like.

A Lincoln sheep in full fleece is an impressive creature, stately, dignified and almost completely covered in long ring-lets of fine wool that brush the ground on all sides. It vies with the Cotswold to produce just about the heaviest fleece of all the Longwool breeds, averaging about 14–16 lb from mature sheep, and from sheep at their first shearing about 26–7 lb. In 2005 a 27-month-old ram, at its second shearing, produced a fleece of 47.5 lb (over 22 kg). This is one of the heaviest and highest-quality fleeces produced by any sheep in Britain, or for that matter the world. Even their foreheads are covered in woolly locks that hang down over their eyes, with only their blue-white ears and noses poking out. The wool is the same quality all over their body, whereas with most other breeds it varies depending on which part it comes from – neck and belly wool being of poorer quality than the rest.

Although they are big sheep, like most of the old English breeds of domestic livestock, they are docile and easy to work with. But their docility can be their undoing. If they are packed too tightly in a wagon or enclosed space, and one goes down, it will suffocate. They just seem to lose the will to live, and quietly accept their fate. They are fine converters into wool and flesh of grass and the residues from crops, such as sugar beet pulp and linseed cake (the residue from linen manufac-ture). Long adaptation to the fens of their native county has also given them excellent feet, with strong resistance to foot rot – an essential attribute for heavy sheep on soft ground. Lincoln lambs are born covered in 'yolk' – a thick yellow sticky protective covering of lanolin – which is a secondary charac-teristic that indicates fine breeding and a gene for good wool.

English wool growing, for manufacture, began when the Romans brought a longwoolled type of sheep into England at some time after they colonised Britain in 55 BC. These sheep almost certainly came from Italy and would have been so different from anything the native Celts kept that they must have seemed to them like mythical creatures, embodying all the grandeur that was Rome. More than twice the size of the Celts' little animals, they had a pure white fleece that was four times heavier than that of the natives' multi-coloured native sheep,

Although the Celts were considerable pastoralists, with large flocks of the type of sheep we met in the last chapter, they were essentially subsistence farmers and their sheep were mostly not kept for profit. Rather, as with all pastoral people, they were a store of energy to see them through the unproductive winter months. For unlike pigs and cattle, which need daily feeding through the winter, sheep can survive on extensive pasture. They are the most efficient extractors of energy from natural vegetation of all the ruminant animals, even goats and deer, and as long as they can fill their bellies they will live through most winters, even snowy ones, surviving on what often amounts to little more than dead grass.

By the time the Romans came, the land had already been extensively grazed over many centuries, probably millennia, by surprisingly large flocks of sheep; the tree cover would have disappeared from much of the uplands and the conquerors would have found a land ideally suited to sheep-rearing. For, in most years, the British Isles enjoys a happy combination of mild winters and cool summers, with no great climatic variation, so that there is usually grazing available throughout the year. In this, Britain, particularly England, is unique. Seasonal migration was largely unnecessary, and even where it was practised, in Wales and the Scottish Borders, it was over relatively short distances. But in many places in Europe, which are either too hot and dry in summer to grow grass or too cold and dry in winter, or both, transhumance was the only way the flocks could be sustained. This involved walking them twice a year, often over long distances, between the winter pastures in the lowlands and the summer pastures in the hills.

The Romans' flocks had been selectively bred since at least the second century BC to produce high-quality white wool for the Empire. Columella, writing about AD 60, refers to three types of Roman wool-producing sheep: the best fleeces came from 'tall sheep' grazing rich flat country in Apulia; 'square-built' sheep came from harder hilly regions; and small sheep from the woods and mountains. The wool from these flocks provided the raw material for the considerable Florentine textile manufacture.

In England, the Romans' wool flocks became concentrated largely around their principal centres of population: Lindum (Lincoln), Camulodunum (Colchester), Gloucestershire around Cirencester, in Kent (Romney Marsh) and in the Midlands. As these places are still connected with the English Longwools, it is tempting to assume they are all descended from a common Roman ancestor. We do not know what the Roman sheep looked like because there are no illustrations before the eighteenth century of sheep with the long, lustrous fleeces of the Longwools, which are so different from all our other breeds. Also, at some point, the Longwools must have acquired their lustre-wool, but we do not know when. It remains a mystery. Nevertheless, it is probably safe to assume that the Lincoln and all the other Longwools descend from those white-faced, polled 'tall sheep' from Apulia that Columella wrote about.

At first, much of the wool from the Roman sheep in England was exported to their factories in northern Gaul (Flanders), but by at least AD 50 they had established a manufactory at Winchester, making garments for the civilian population and the army. The finest English wool was much in demand from the highest ranks of Roman society, because it produced garments of the best quality, and the value of his garments proclaimed a Roman's social status. Coarser wool went to make things like socks for the infantry to wear inside their hob-nailed marching sandals.

After the Romans left, the English wool trade did not disappear. English flocks continued to produce high-quality wool almost irrespective of who was the dominant power. There is a surviving letter written in 796, from Charlemagne to King Offa of Mercia, in which is mentioned the supply of English woollen cloaks, or it could be cloths – the document is not clear. The Saxons took a great interest in the growing and manufacture of English wool. Edward the Elder's first wife, Egwina, was proudly described as a shepherd's daughter, although she was really the daughter of a country gentleman, or a yeoman farmer, as opposed to a soldier. He was a 'shepherd' because those who were not devoted to the military life were employed in the life of the land, of which the most significant activity was the keeping of sheep.

Wool played a pivotal role at all levels of Saxon society. King Edward 'sette his sons to scole, and his daughters he set to woll werke, taking example of Charles the Conquesterer'. The very name by which unmarried women in England are still designated, *spinsters*, is proof of the antiquity and importance of wool-working and how central sheep were to English national life and prosperity. The flocks must have been numerous and large to have given employment 'to the unmarried women of all classes, from the daughter of a prince to the meanest person'.

> *To spin with art in ancient times was seen,*
> *Thought not beneath the noble dame or queen:*
> *From that employ our maidens took the name*
> *Of spinsters, which the moderns never claim.*
>
> Youatt on Sheep, p. 196

Then after the Norman Conquest, and for about three centuries during the Middle Ages, English wool growing reached its apogee. Wool became so valuable that fortunes were made from growing it, dealing in it and manufacturing it and an unknown merchant could engrave into the windows of his house, *'I praise God and ever shall – it is the sheep hath paid for all.'*

In 1189 King Richard, Coeur-de-Lion, succeeded to the English throne and almost immediately embarked on the Third Crusade. On his way back from Palestine in 1193 he fell into the hands of Henry VI, the Holy Roman Emperor, who held him for the exorbitant ransom of 150,000 marks, which was 75,000 oz, just over two tons of silver and the equivalent of three years' income for the English Crown. The kingdom was scoured to raise the vast sum. Both laymen and clergy were required to give up a quarter of their income for that year. The Church was expected to donate most of its gold and silver plate and the Cistercian monastic houses were asked for a year's clip of wool. Thus arose the saying that Richard the Lionheart's ransom was paid in English wool.

Throughout the Middle Ages long-staple wool from Lincolnshire and the Midlands was only matched in quality by the short-staple wool from the Welsh Marches. Staple is a measure of the length and fineness of wool fibres. Francesco Pegolotti, an agent for the Italian merchant and banking house of Bardi in Florence, travelled throughout England during the early part of the fourteenth century buying wool. In his merchant's handbook, *Pratica della mercatura*, he recorded the prices for English wool between 1317 and 1321. The highest priced was the 'Lemster Ore', which came from sheep that grazed the lands of the Cistercian houses of Tintern Abbey and Abbey Dore. This was some of the finest wool in Europe. Queen Elizabeth I insisted that her hose be made only from Lemster Ore, refusing to wear any other. The sixteenth-century bucolic poet Michael Drayton celebrated its virtues:

Where lives a man so dull, on Britain's farthest shore
To whom did never sound the name of Lemster Ore
That with the silkworm's web for smallness doth compare.

Lemster Ore was worth 28 marks per sack of 364 lb. A mark was two-thirds of a pound – 13s 4d (66.6p) – but no mark coins were ever struck because a mark was always worth 8 oz of silver. At today's silver price of £12.50 an ounce (approximately £200 a pound), 28 marks would be the equivalent of £3,000, whereas similar modern Shortwool fleeces would now be worth about £220 a sack.

The Lemster Ore probably did not come from any uniform type or breed, but was created by the land the sheep grazed and the way the flocks were kept. The old type of sheep which gave the Lemster Ore was an ancient English shortwoolled heath sheep which, at some remote time, had occupied all the English and Welsh sheep-walks, but was pushed further into the west by the Saxon invasions. It was a small, fine-boned, very hardy animal with an exceptional ability to thrive on scanty pasture, and 'deserved a niche in the temple of famine'. It was the sparseness of the unimproved ancient pastures that made the wool fibres so fine. When such sheep are grazed on lusher pasture, their wool deteriorates and becomes coarser.

This, coupled with unwise crossing with modern breeds, particularly the New Leicester, in an attempt to improve the carcase, damaged the wool even more. Flockmasters discovered (often too late to repair the damage) in their zeal to grow a better carcase that improving the carcase reduces the quality of the wool. The modern descendant of the old type is supposed to be the Ryeland, which is a medium-sized solid square sheep, naturally polled in both sexes and covered in short, dense, fine wool all over its face and down its legs almost to its hooves. But the wool of the modern Ryeland is so much inferior to that of its forebears, having been ruined by injudicious crossing, that it is but a shadow of the sheep whose fleeces once rivalled the best of Spain. Its modern manifestation is the recherché interest of smallholders and sentimentalists, and a woolly relic from a vanished age.

There is a stunning list showing the price range of wool in 1343 in marks per sack from different counties of England. We can see that the highest-priced wool came from a few counties that run in a swathe from the Lincolnshire Wolds through Leicestershire into Staffordshire, the Welsh Marches and across into the Cotswolds. The highest price was 14 marks (£1,400 today) per sack, for the long-staple wool from Lincolnshire and Leicestershire and for the short-staple wool from the Welsh Marches. By comparison, the 2012/13 price for lustre wool (of similar quality to Lincoln) is 380p per kilo, or just over £600 a sack.

From ancient times, wool had its own special weights, which were used well into the last century:

7 lb = 1 clove
2 cloves = 1 stone (14 lb)
2 stones = 1 tod (28 lb)
6½ tods = 1 wey (182 lb)
2 weys = 1 sack (364 lb)
12 sacks = 1 last (nearly 2 tons)
a 'wool-pack' was 240 lb

Wool buyers and farmers would have kept two clove (half-stone) weights, often in the shape of a shield, and certified as accurate by the Crown, and stamped with the Royal Arms. The two weights were pierced and held together with a leather strap. When the wool buyer came to negotiate for their wool, farmers would put the buyer's two 7 lb weights on one side of a scale and a stone on the other from which they would chip pieces until it weighed exactly 14 lb, which is probably the origin of a *stone* being 14 lb. The oldest known weight is a pear-shaped stone wool-weight, carved from limestone, pierced at the top and weighing 1.5 lb. It came from Lagash in Sumer and bears a cuneiform inscription: 'Mina for wool issue, Dudu high priest.' A *mina* was equal to 60 shekels and Dudu was high priest of Lagash in Sumer in about 2500 BC. (The stone is now in the Ashmolean Museum in Oxford.)

For centuries much English wool was exported to weavers in Europe, primarily to Flanders and Italy, who were still pre-eminent in its manufacture more than 1,000 years after the Romans had established the industry, and are still important today. It was valued for its fineness and smoothness, or *handle*, particularly the long-staple wools that came from the very sheep the Romans had introduced all those centuries earlier. Land-owning dynasties were founded on wool and the English Crown drew a large part of its revenue from it. Wool was the North Sea oil of its day, except its bounty lasted for four centuries rather than the four decades of oil. As an example of just how much the medieval Crown depended on its income from wool, in the last year of Henry V's reign (1421–2), out of a total income of £55,750, £35,000 came directly or indirectly from wool.

Wool growing created the landscape of Britain and, in a magnificent medieval symbol of the pre-eminence of the wool trade in England's fortune, there are two Woolsacks in the House of Lords. One was occupied by the Lord Chancellor and a larger Judges' Woolsack by the Law Lords. The Lord Chancellor occupied his from the fourteenth century, when Edward III commanded that his Chancellor should sit in the House of Lords on a sack stuffed with wool, until July 2006, when the Woolsack became the seat of the Lord Speaker as the presiding officer in the House of Lords. Both Woolsacks are large cushions covered with red cloth and stuffed with wool. The Speaker's cushion has a back-rest in the middle but no back or arms. In 1938 it was found to contain horsehair and was re-stuffed with wool from all parts of the Commonwealth, as a symbol of unity.

Until the middle of the fourteenth century the profit from wool growing accrued disproportionately to the feudal land-owners. Most land was subsumed into the feudal pyramid and occupied by villein tenants with personal obligations to their superior lord. There were huge flocks of sheep grazing the prehistoric downs and feeding in their tens of thousands on the communal and inter-manorial marsh pastures, but there was limited scope for ordinary people to establish their own com-mercial flocks because most of the manorial land that could have been used for grazing was cultivated.

The mass of villein tenants were also unable to accumu-late significant capital, either in flocks or in land. They were not slaves or serfs; they had legal rights according to their status and carried out their obligations because custom and kinship demanded it rather than under coercion, but the flocks they tended benefited their feudal lord. Things were chang-ing gradually under increasing pressure from a growing cash economy, and the tenants' obligations were slowly being com-muted into money payments, whose payment avoided them having to work personally for their lord. But the changes were slow and patchy, and only affected the edges of a system whose core remained largely unchanged. But within two years of the summer of 1348, all this was changed utterly when the whole fabric of English society was torn apart.

A run of cold wet summers and terrible harvests culmi-nated in the awful summer of 1348. By August grain crops were rotting in the fields, corn sprouted in the ear, beyond hope of harvest, and widespread starvation was predicted. But, as if to prove the old saying that a foreseen famine never happens, the plague that was sweeping its way across Northern Europe was something far worse than famine. The Black Death became a catalyst for the most fundamental change in land use in

England for many centuries. Up to a third of the population of Europe is believed to have died in three years between 1347 and 1350. And in England it has been estimated that 2 million people perished out of a population of about 5–6 million. The infection spread so rapidly that its victims 'took breakfast with their friends and dined with their ancestors in paradise'. Whole villages were abandoned and fields left uncultivated: the Winchester Estate Rolls typically recorded the rent received from numerous tenants as being 'nothing, because he is dead'.

Manpower was so scarce that the laborious manorial cultivation of the early Middle Ages became impossible. For the first time in history, large acreages of arable land that had been cultivated for many centuries became grassland, colonised by the grasses and wild flowers that grow naturally in the English climate. This proved to be a godsend to sheep keepers. The great value of grass is that it grows from the root, not the tip, so removing the leaf does not retard its growth; rather it causes it to tiller out and produce more stems and leaves. And as it is grazed, grassland gradually improves. Weeds are suppressed because removing their leaves kills them and the manure from grazing sheep and their treading increase soil fertility with every passing year. This grazing and treading by sheep is 'the hoof that turns sand into gold'. The most beneficial grazing is with mixed cattle and sheep because sheep bite and nibble, whereas cattle tear at the grass, removing old and weak stems, and in combination this keeps the sward young and healthy.

Very quickly the abandoned manorial fields became productive permanent pasture, which required little maintenance other than looking after the flock that grazed them. On many of the heavy clay soils, particularly in the Midlands, permanent pasture was a better way of farming, because ploughing and cultivating such land is a tricky business that must be done when the soil is exactly right: too wet and it's too sticky to work and too dry it sets like concrete.

In the ensuing century and a half great fortunes were made from keeping sheep on land that had fallen into permanent pasture. In the fifteenth and sixteenth centuries, for example, in the Midlands, the Spencers made their fortune from wool growing. They were by no means the only ones, but they were among the most successful. By the end of the sixteenth century Robert, first Lord Spencer (1570–1627), had £8,000 a year from 20,000 sheep in the Midlands at a time when a head-shepherd (a position of some authority) earned about one mark a year – half a pound of silver worth about £150 at the value of silver today.

Even by the middle of the seventeenth century wool was still the most valuable thing that sheep produced. In 1641 Henry Best, a yeoman farmer from Elmswell near Driffield in the East Riding, wrote *Rural Economy in Yorkshire*. Best's father was a member of the rising Elizabethan mercantile class who had bought his little estate with money he made in the City. In 1641 Best sold his entire wool-clip for the year, 29 stones (406 lb or 184 kg) for £11 12s, and in the same year he records that he hired a man for £3 for the year. Louise Fairburn's thirty breeding ewes clip an average of a stone of wool each (14 lb) – 30 stones from the flock; this is the same weight as Henry Best's was 400 years ago. But, in 2012, she got £240 for all the wool from her flock and she would need a hundred times that just to employ one man.

Soon after the Conquest the English Crown had set up the Wool Staple (from the old French *estaple*, meaning open market). This monopoly levied duty on all raw wool exported by requiring that it be sold only through depots controlled by the Staple. In the ensuing centuries much of the Crown's revenue came from duty exacted by the Staple on the export of English wool to Flanders. (Interestingly, we still export more wool to Flanders than to any other country in the world.) This was why the Wool Staple was run from Antwerp until 1353, when Edward III brought it home and temporarily established it in fifteen depots in towns across the kingdom. Then in 1363 he moved it to Calais, where twenty-six English merchants formed the Merchant Staplers Company. They grew wealthy from this monopoly, and were protected by the English Crown, which in return expected to be entitled to borrow money from the Staplers whenever it needed it. This suited the Crown because it could avoid the inevitable concessions that Parliament would have exacted in return for granting money. The Staplers' monopoly and the taxes imposed raised the price of woollen cloth in Europe. The combination of the resulting hardship to the Flemish manufacturers and the inducements offered by the Crown to settle in England was an offer they could not refuse. They moved to set up in business in England, which invigorated English cloth manufacture, reduced the export of raw wool and enhanced the revenues to the Crown from the increased trade and duty on manufactured goods.

The British Wool Marketing Board is the modern incarnation of the Staple, with a statutory monopoly over the sale of British wool. The difference is that its role is to get the best price for the producer rather than collect revenue for the Crown. Its head office and warehouse are in Bradford, a city in a county whose very names are synonymous with British wool. The UK annually produces over 22,000 tonnes of wool from about 23 million sheep. Anybody who keeps more than four sheep is obliged to sell their wool to the Board, apart from Shetland producers, who negotiated an exemption. Most British wool is currently exported, because the indigenous textile manufacture has taken a hammering in recent years from foreign competition, notably from China and India, because its profitability is directly related to the cost of labour and the strength of the currency. Currently Britain is down and others are up (for obvious reasons) but there are signs that this is changing.

All British fleeces arrive into the BWMB's Bradford

warehouse (or one of its regional depots) from the farm in 'sheets' – big sacks about six feet square, each packed with about fifty fleeces. Traditionally the sheets were made from woven jute fibre, but now they are white polypropylene because it is more durable. No wonder wool is losing out to man-made fibres when even the Wool Board eschews natural fibre in favour of man-made. Nearly all the annual clip of wool is sold at one of the twenty-two fortnightly auctions held in Bradford. These are the quietest auctions I've ever been to: there is no auctioneer, the bidding is done by the click of a mouse at a computer terminal and the bids and final price are recorded on a big screen at the end of the room.

The Board pays the producer a guaranteed price based on an average of two years' auction prices. There are over 115 grades of wool in the Board's price schedule, and the grader's skill, learned over a long apprenticeship, is to sort each fleece into its proper grade by hand. The fleeces have been wrapped on the farm (there is a penalty for unwrapped fleeces) by folding the outside edges into the middle, then rolling the fleece from the tail end and twisting the neck wool into a band which is tied round the rolled fleece. The fleeces are tipped out of the sheets onto big tables and the grader goes through them, sorting them into their appropriate grade, and throwing each into the appropriate bin behind him. When the bin is full its contents are mechanically squeezed into bales which are then wrapped in orange plastic. A mechanical corer is then plunged into the bale to take a sample which the Board uses to assess and guarantee the quality of the wool in that bale.

Wool is a unique and complicated substance. Each fibre consists of a protective layer of overlapping cuticle cells that lie, like the slates on a roof, towards the tip. Each cell has a waxy coating that repels water away from the root, but allows water vapour to penetrate and be absorbed by the fibre. That is why wool can absorb up to 30 per cent of its weight in water and not lose its insulating properties, and when removed from the sheep can absorb a large volume of dye. It is flexible, fireproof and resistant to stretching and folding, so it is simply the most versatile and valuable natural fibre we have. It has been much underrated in recent times, losing out to cotton in the eighteenth and nineteenth centuries and to man-made fibres in the twentieth, but nothing can really replace wool. And Lincoln wool is some of the best.

The global pre-eminence of English wool is reflected in the worldwide acceptance in the wool trade of longstanding English terms and standards. English wool graders devised the Bradford Count, a measure of assessing the fineness of wool. It is an estimate (relying on the skill of the grader) of the number of hanks (a hank is 560 yards) of single-strand yarn that can be spun from a pound of 'top', which is washed wool combed to make all the fibres lie parallel. A pound of top with a count of 56 could be spun into 56 hanks of 560 yards each, which is 17.8 miles of yarn. The Bradford Count was used across the world for centuries, until it was superseded by measurement by microscope expressed in microns (0.001 mm). A Bradford Count of 56 would now be 28 microns.

A staple, another English term in common use, is a lock of wool, and staple length determines how easy it is to spin into a yarn; wool with shorter staple lengths is used for woollen manufacture, such as knitwear, and longer staple lengths for worsted cloths for suiting. Staple strength is the force required to break apart a lock of wool by pulling it at the ends. This tells the manufacturer how well the fibres will cling together in spinning. It has its modern measurement in newtons per kilotex. The quality of the wool is also determined by the place along the staple where it will break if force is applied to try to pull it apart: at the base, in the middle or at the tip of the lock.

Crimp is the number of waves along the length of each fibre, which is reflected in the waves along the length of each staple. Finer fibres with a high Bradford Count usually have a tighter crimp with more waves per centimetre of their length and are not suitable for worsted cloths because they are not easy to flatten out, so they are mostly used for high quality woollen knitwear. Crimp is caused by each fibre having cells of two different types arranged into two distinct groups, one on each side of the fibre, and one group expanding further than the other and bending the fibre. Its natural function is to add bulk to a fleece to trap a greater volume of air between the fibres and give the sheep greater insulation from the elements.

At its height, English wool was nearly as valuable as the finest Merino wool from Spain, with the difference that the best English wool was longer-stapled (and probably lustred) whereas the Merino was like the Lemster Ore, short-stapled with fibres nearly as fine as silk. The best long-staple English lustre wool came from around Lincoln and Stamford. And the best and most lustrous of all came from the sheep that grazed the coastal plain between Boston and Louth. It is not known why this should be. One (rather fanciful) theory is that it was the effect on the fleece of the moist misty air and heavy morning dews laden with sea-salt.

The lustre or natural sheen of these wools is caused by the scales of each fibre turning inwards and presenting a smooth, shiny and more rounded edge to the surface. Lustre is a feature of a few high-quality wools and gives the sheen to worsted cloth. Lustre wools take dyes readily and give strong clear colours to the cloth. Worsted cloth is named after the Norfolk village of Worstead, and is made by a spinning process that provides tight and smooth fabric which recovers quickly from creasing. That is why lustre wools are more valuable than ordinary wool.

Fulling scours and mills the cloth in one operation: scouring removes the oils and dirt in an alkaline solution, and milling shrinks and firms it. Felting is an extreme form of milling by which the cloth is beaten into a kind of mat by intensifying the natural propensity of the fibres to cling together. The Romans scoured their cloth by immersing the bolts in vats of stale human urine (which was carefully preserved for the process) and then having slaves tread them. Not surprisingly by medieval times fuller's earth was preferred. This is a type of clay with the power to absorb the lanolin and cleanse the wool of impurities. The difference between worsteds and wool lens, which obviates the need for fulling, is that the long-staple wools used in worsted fabrics are combed to make the fibres lie parallel to each other in the same direction – 'butt end to tip', i.e. from the end that was cut during shearing to the tip of each fibre. The process removes the wool's natural crimp, straightens it and causes the fibres to cling together.

Between the 1850s and the 1920s, when the Lincoln was still a force in the land, many thousands were exported for cross-breeding to create the sheep of the New World. The flocks of South America, Australia and New Zealand were bred from the Lincoln, and then during the twentieth century Argentina and Russia also bought rams to improve their wool flocks. In New Zealand they went to create the Polwarth, a dual-purpose wool and meat breed, one-quarter Lincoln and three-quarters Spanish Merino, and the Corriedale, a quarter Merino and three-quarters Lincoln. There are now estimated to be 100,000,000 of these two breeds grazing the southern hemisphere.

But by 1971 fewer than 500 breeding Lincoln ewes remained in fifteen small flocks. Numbers increased a little in the 1980s and have stabilised at about 700, most kept for sentimental reasons. There is a minor upsurge of interest in the breed from hobby farmers that may well ensure its survival, but not as a commercial proposition unless wool becomes much more valuable. It is poignant that a breed which for more than a thousand years grew a large part of the wool that made

England rich, and provided one half of the genes that created two of the most successful wool-producing breeds in the New World, should be so diminished.

During the second half of the last century wool became a nuisance. Its value was so low, for so long, that most British sheep farmers could not have cared less what happened to their wool after it left the farm. No longer would the wool cheque pay the farm rent. Farmers just needed their sheep to be clipped as quickly as possible, largely for the good of the sheep. Things have begun to look up slightly in the last few years, but the recent modest increases in price have to be compared with their lowest point, when it cost more to shear most sheep than their wool was worth. Farmers took to burning it or burying it rather than losing money by paying to have it hauled to their nearest wool depot. Recently, in another British initiative, the Campaign for Wool, whose patron is Prince Charles, has done much to promote wool worldwide; coupled with a reduction in supply because the world's sheep flock has been reduced, the market price has risen about threefold, but wool is still an insignificant part of a sheep farmer's income.

But all this wool from big sheep, like the Lincoln, comes at a price. They can carry so much wool, especially if they are heavily in lamb, that it can make them top-heavy and vulnerable to being cast, i.e. rolling onto their backs and being unable to stand up again. When this happens the fermenting vegetable matter in the rumen, the largest stomach, blocks the outlet of gas to the oesophagus, it builds up, the rumen swells, the heart stops and they quickly die. They are also susceptible to fly-strike because, when they eat lush spring and summer grass their wool can become soiled by scour (diarrhoea to you and me). It is easy to spot a dirty fleece on a shortwoolled sheep, but heavy-woolled Lincolns, in warm damp weather, have to be inspected regularly, because the fleece can appear clean on the outside, and yet hide a multitude of horrors beneath.

The fly strikes silently, its eggs hatching into maggots deep in the wool, without any outward sign. The remedy for this is dagging. When the fly is most active in early summer, all soiled wool must be clipped away. It is a dirty, back-breaking job that needs great care with the shears, because flies are nearly as fond of wounds as they are of muck.

The blue- or greenbottle 'fly' is attracted by the smell of decay. Leave a piece of meat out of the fridge for a few hours in summer and you will see how quickly tiny white fly's eggs appear on its surface in clusters. It is the same with a dirty fleece. Little blind creamy maggots soon hatch out and burrow into the moist areas, where they immediately begin to gorge themselves first on dead, then on living flesh, and if they are not stopped they will eat the animal alive. Once the feeding starts, the smell from decaying flesh and maggots attracts more flies, which lay more eggs that hatch into more maggots and the animal is quickly overwhelmed.

During the early stages of fly-strike the afflicted sheep becomes mildly restless, sitting down and standing up and moving around pointlessly. Within a day or so it becomes more uneasy, sometimes shaking its legs in a sort of involuntary dance, stretching its neck backwards and nibbling frenziedly at its hindquarters. Very soon it will be unable to lie or stand still, moving around compulsively, regularly emitting moans and noises, as it is tortured by the maggots. Sometimes it will try to bite at them where they are feeding, and show signs that it is being tormented by the pain. In the terminal stages of infestation the animal's will is broken and it leaves the flock to hide away and die.

Sheep can also get maggots in their feet, attracted by the smell of decay caused by foot rot. At first this can be beneficial because they will eat away the rot, but if they aren't caught in time they move on to the living flesh. Fly-strike is one of the nastiest things a shepherd has to deal with, although its worst excesses are controlled by modern dipping. I once came across a young sheep that was being pestered by swarms of bluebottles. It hadn't the energy to run off and I immediately saw why. The fleece attached to a large strip of putrid skin around its hindquarters and along its back and flanks came away in my hand to reveal a seething mass of maggots gorging on its flesh. They wriggled out of its anus and vulva, which they had partly eaten away, and dropped bloated to the ground, sated on their living victim. The sheep moaned and strained, and hundreds more poured out, blindly burrowing into the grass. The stink of putrefaction had to be smelled to be believed. I retched as I dragged the doomed creature to the nearest tree to tie it, while I ran home to get my gun.

If wool breeds like the Lincoln are to survive in the modern meat-producing world it is ladies with a paddock, like Louise Fairburn, who will keep them going. The Lincoln and all the other wool breeds that commercial farmers can no longer afford to keep would not last five years without them.

In their native Lincolnshire, Lincoln sheep seem at one with their landscape – their creamy wool tones perfectly with the limestone walls. It might just have been in my imagination, but these sheep seem to be part of the soil upon which they have lived for nearly two millennia and have taken on its hues. I also fancied I could see something of the ancient Roman lineage in their proud bearing and a gentle grace that out-classes less noble, utilitarian breeds.

✌ 3 ✌

THE NEW LEICESTER

My people want fat mutton and I give it to them ... Sheep for the keelmen,
pitmen and all such hardworking people are never too fat for people of these
descriptions.

Robert Bakewell

OST OF US AT SOME TIME OR ANOTHER WILL
have come across a couple of rather unprepos-
sessing-looking parents who seem to have bred
a good-looking daughter. She bears so little resemblance to
either of them that we are driven to wonder where on earth
she came from. An uncharitable observer might speculate that
her mother had not been entirely faithful to her father, or that
some mix-up could have occurred in the maternity unit. But
then, on closer inspection, we might notice that some of her
more pleasing features could have come from her mother: her
nose, her hair, her ankles or whatever. Then if we look a little
more carefully her eyes might have come from her father, and
the tone of her skin and the way she holds herself. Then we
notice that the reason she is so much more attractive than
either of her parents is that she has inherited the best features
of each and the worst are nowhere to be seen.

That is the effect that Robert Bakewell's New (or Dishley) Leicester sheep had on nearly every breed it was crossed with. The result was startling – hence the sobriquet 'The Great Improver'. It did not improve *every* aspect of *every* type it was crossed with; some, such as the Lincoln and Ryeland, whose wool deteriorated, were badly affected by the match and never recovered from the infusion of Leicester blood. But it always had a striking effect, particularly in the carcase, and when its effect was beneficial, it imbued its offspring with just those characteristics that were needed to bring them into the modern age.

As the eighteenth century opened, sheep farming operated largely as it had done for over 500 years; wool and tallow (for candles) were still the most valuable part of the sheep, with its manure an aid to soil fertility, particularly on thinner soils. Distinct breeds, as we know them today, had not yet been developed; rather there were types which had become adapted to the soil they lived on by long association and had changed little over the previous centuries. Some were Longwools and, like the old Teeswater, by today's standards huge ungainly carcases of mutton clothed in as much long-staple wool as they could carry; others were smaller and grew shorter wool; some were white-faced, others black-faced, some horned in both sexes, others only horned in the rams, and some were polled in both sexes. But the thing they nearly all had in common was that they were not meant to be eaten, or at least not when young. It was considered wasteful to eat animals that were barely grown and immature, especially before they had had the opportunity to give a fleece of wool. And by the time they were mature enough to eat, their meat was often unpalatable, stringy and tough, coming as it did from three- or even four-year-old animals.

These types of sheep had served well enough for many centuries, to satisfy the market for wool at home and abroad, but they proved wanting when faced with the growing demand for meat from a newly urbanised population that needed somebody else to grow their food for them. And the transformation that ensued was, in its way, as far-reaching in the eighteenth century as the development of the internet and the telecommunications revolution have proved in the twenty-first.

Farming, the most traditional of occupations, which works with the unvarying cycle of the seasons, does not lend itself easily to radicalism. In earlier times, most farmers resisted change because one bad decision could lead to ruin and the destitution that was never far away. A farmer would either have to have been a madman or a prophet to risk going against the wisdom of the centuries. That is why many eschewed, or probably were never aware of, the opportunities for profit that beckoned to those who could give the people what they wanted. It was left to a handful of farsighted pioneers, during a few short decades in the middle of the eighteenth century, to recognise that they stood on the brink of a revolution.

Into this world of opportunity emerged Robert Bakewell of Dishley Grange in Leicestershire, who was probably a bit of a madman and certainly a lot of a prophet. He was born on 23 May 1725 and while still a young man began the experiments on his father's farm that so radically changed the face of livestock breeding right across the world. Before the emergence of men like Bakewell, medieval stockbreeding had been a haphazard affair, described by one writer as the union of nobody's son with anybody's daughter. Bakewell was probably not the originator of the new breed, but he refined and brought to the pitch of success the transformation that others had already begun, notably Joseph Allom, another remarkable Leicestershire man, somewhat overlooked in the clamour over Bakewell's achievement, who deserves considerable credit in that he 'raised himself by dint of industry from a plowboy, and thereafter raised the old Leicester sheep from its medieval mediocrity'.

Bakewell took over the tenancy of Dishley Grange after his father's death in 1760, and made its name synonymous with his new breed of sheep, the Dishley (or New) Leicester. His achievement was to create a breed from the old Leicester Longwool that became the fulcrum between the ancient wool-producing types and all the modern meat breeds. There is not a breed of sheep in the industrial societies of the Western world that does not have at least a little of the blood of Bakewell's Dishley Leicester running in its veins.

During a few years in the middle of the eighteenth century Bakewell took this slow-to-mature, long, weak-in-the-frame old Leicester, described at the time as 'a loose and irregular, slow-feeding, coarse-grained carcase of mutton', and transformed it into a quick-growing, hornless, short-legged, barrel-shaped animal with a greater proportion of meat to bone than nearly any other type. He reduced the inedible parts of the carcase – the bone in particular – and in so doing produced more meat from each animal for the least consumption of food, in the shortest possible time. And surprisingly, despite its being created within the course of only a couple of decades, the new sheep bred true to type; in other words, its offspring had the same characteristics as their parents. Bakewell's insight was that he 'perfectly understood the relation which exists between the external form of an animal and its aptitude to become fat in a short time'. His sheep were described at the time as being always 'semi-fat', i.e. they carried a covering of flesh with a thin layer of fat over it, and therefore could be finished for slaughter in a much shorter time than other breeds.

'Fat' in this context needs some explanation. It is nearly synonymous with 'mature' and means that the animal has reached a stage of growth at which it has developed all the muscle that it ever will and has begun to lay down a covering of fat over its carcase and within its muscles. It is very far from the popular understanding of fat as being obese. A fat sheep is more like a lean healthy athlete with well-formed muscles. Bakewell sought to breed a sheep that became fat, i.e. matured, at a younger age than the great, donkey-like, rangy Longwool sheep, and with less bone and offal and with smaller joints of high-quality meat for the tables of his 'people'. Bakewell memorably declared 'all is useless that is not beef'. By this he meant that the part of a carcase of a grazing animal that farmers ought to be aiming for was the most valuable cuts of meat, not bone or wool or fat.

The traditionalists, particularly the prominent breeders of the old Lincoln sheep, complained that his innovations were ruining their breed's main attribute, its valuable wool. This criticism was not without foundation. But they seemed unable to foresee the tremendous increase in urban demand for meat and that the wool was an inevitable sacrifice. Bakewell was prepared to admit that his emphasis on carcase was damaging the wool of the breed that had once made England rich, but he thought it a price worth paying, as is shown by his reply to the Lincoln breeders' criticism: 'it is impossible for sheep to produce mutton and wool in equal ratio: by strict attention to the one, you must, in a great degree, let go the other.'

Of the Dishley Leicester's many other detractors, some were motivated by jealousy and ignorance, but there were others whose criticisms were valid so far as they went: its flesh *was* coarse and its stocky build sometimes made lambing difficult; it *did* tend to run too early to fat and *was* heavier in the front end than the rear – where the dearer cuts are found. But Bakewell replied to them all with his characteristic brio that 'mutton *anywhere* was welcome to the poorer classes', and that fat mutton found 'a ready market amongst the manufacturing and laborious part of the community'. Because of all these deficiencies, as a pure breed in its own right, the Dishley Leicester proved disappointing.

Another famous agricultural improver, the aristocratic

Thomas Coke from Holkham, in Norfolk, began his own improvements by keeping Dishley Leicesters, and sold their mutton 'at 5d the pound to poor people in Norwich'. But he would not eat 'the insipid fat meat' they produced. He preferred the sweet lean mutton of the ancient Norfolk breed from the Brecklands, notwithstanding his decades-long (eventually successful) campaign to eradicate Norfolks from his estate and from the county. Coke subsequently disavowed his Dishley Leicesters in favour of the improved Southdown created by yet another great eighteenth-century improver, John Ellman from Sussex.

But the Dishley Leicester's peerless value turned out to lie in its quality as an improving breed. This characteristic made it so popular that by the end of the eighteenth century it had spread throughout Europe and during the early part of the nineteenth its descendants played a crucial role in sheeping the New World. It was also found to possess the remarkable property of encouraging milk production and fecundity in its cross-bred daughters and of being able to breed with smaller ewes without causing them to have over-large lambs.

Bakewell was a remarkable man in an age that made remarkable men. He was a self-confident, even cussed character, described by one who knew him well as 'a tall broad-shouldered, stout man of brown-red complexion, clad in a loose brown top coat, scarlet waistcoat, leather breeches, and top boots'. Sir Richard Phillips, who made engravings of Bakewell's sheep, knew him well and compared him with William Cobbett:

> In originality and in self-thinking on every subject I have never met with any approaching to him except in William Cobbett ... in his opinions he was as bold and original as Cobbett. The vulgar farmers hated him, just as the small thinkers hate Cobbett ... there was the same playfulness

of manner, the same contempt of vulgar opinions and of authority in thinking, and the same deviation from common tracks both in conclusions and practice...

He was a bachelor, who did everything by the light of reason and appeared to have no time for marriage, perhaps because he felt it might have distracted him from his great project. His household was presided over by his sister, who ran Dishley Grange as her domain, where she kept almost open house for all who turned up to learn from the great man during the three decades when he was at the height of his success. Such was his fame that Prince Grigory Potemkin sought his opinion on agricultural matters, and Catherine the Great sent seven or eight young Russian men to be his pupils, with a view to establishing an imperial farm; but the empress lost interest and the venture came to nothing. All his guests, irrespective of rank, whether the sons of Russian noblemen, or tenant farmers from Yorkshire, ate side by side at a long table in his kitchen. Bakewell habitually took his meals by himself at a small side-table and was such a creature of habit that he never allowed his guests to interfere with his strict daily routine. No matter who was staying, he always retired at eleven, 'having knocked out his last pipe at 10.15'.

It is astonishing what Bakewell achieved by instinct, coupled with powerful determination to follow it, against long-settled custom and initial hostility. He achieved rapid results by ignoring the almost unbreakable taboo against incest, practising 'in-and-in breeding', or 'line breeding' as it is called nowadays. His method was to choose animals with characteristics nearest to those he was looking for and then breed them together. It made no difference whether or not they were close relatives. If he considered it beneficial and necessary, he bred mother with son, father with daughter – and repeated the process, ruthlessly culling any defective offspring, until he had obtained the desired result.

At the time it was an outrageous thing to mate his Long-
horn cow 'Old Comely' with her son 'Twopenny' (so named
because a farmer who saw the calf remarked that he wouldn't
give twopence for him). His crossings might well have produced some odd offspring, but if they did, there is no record
of them and, in his careful, secretive way, he would have dis-
posed of them and kept quiet about it. He never scrupled to
cull any animal that didn't come up to his exacting standards.
There had long been a taboo on close-breeding of animals
which followed the Bible and the Church's injunction against
the practice for humans. In George Culley's *Observations on Live
Stock*, 1786, he writes about the 'common and prevailing idea
... that [no ram] should be used in the same stock more than
two years ... otherwise the breed will be too near akin. Some
have imbibed the prejudice so far as to think it irreligious.'

It was not *some* who had *imbibed the prejudice*; it was an
almost universal taboo considered contrary to the law of
God and of nature. Bakewell dared to break it and in so doing
achieved results that discomfited his rivals. Many really did
think what he was doing was immoral, but some used their
moral disapproval as a cloak for their jealousy of his success. He
was convinced that the ends justified the means, which was not
something early-eighteenth-century traditional society found
easy to accept. But by the end of the century his methods had
become so widely accepted that people had forgotten things
had ever been any different, and probably forgotten the name
of the man who had caused such a revolution.

Charles Darwin credited Bakewell's work as an influence
on his theory of natural selection, saying in *On the Origin of
Species* that the changes brought about by Bakewell's selective breeding demonstrated 'variation under domestication'.
Darwin uses the example of two separate flocks of sheep ('Mr
Burgess's and Mr Buckley's') which came from Bakewell's original stock, and which over fifty years had become so different

from one another that 'they have the appearance of being quite different varieties'. Bakewell achieved his variation under domestication with the old Leicester sheep in less than half that time by his controversial in-and-in breeding.

Bakewell was a Unitarian and it is tempting to connect his pragmatism with his dissenting. In an age in which formal religious practice still pervaded every aspect of national life, dissenting was an expression of rational rebellion against the zeitgeist. Dissenters still went to church or chapel, but they prided themselves on believing that they had thought out their own salvation, and there were few who applied this more assiduously than Bakewell.

He was a great advocate of improving soil fertility by the percolation of flowing water, which he practised in his meadows and pastures. He went to the huge expense, on land of which he was merely the tenant, of digging between ten and twelve miles of canals to carry water from a stream on his land to fertilise his meadows. He also used these watercourses for transporting produce by boat around his farm, particularly floating turnips from where he grew them to a kind of holding place near the farmyard, where they arrived already washed and ready to feed to his housed livestock.

Before it was accepted practice, he subdivided grazing fields into small enclosures by planting quick-thorn hedges so that he could rotate his grazing stock and make more efficient use of the grass and cause it to grow more vigorously. He made his farm roads concave rather than the traditional convex, so that water flowed across the surface of the road and into the middle, to clean and improve it. He grew large acreages of cabbages and turnips for winter feeding of his stock. He grew Dutch Willow and harvested it on a seven-year rotation, using the timber for making all manner of things around the farm, including tool handles, gates, fences, farm implements, and so on. When he was criticised for wasting land with the wide hedges dividing his grazing fields, he replied characteristically (in a note quoted by Arthur Young, first Secretary of the Board of Agriculture, and his great supporter):

> If it be objected, that by this method, there is a waste of land, it is answered, that when the price of coals exceeds sixpence a hundred, the hedges make as much for the fuel as the land is worth for any other purpose.

Perhaps his greatest innovation was the way he evaluated the effect of his rams on a wide range of different types of ewe by hiring them out (as he did with his bulls). When he first tried the practice it was ridiculed by many of his wiseacre farming neighbours and few were prepared to pay him much of a fee to use his rams. Some even thought he ought to be paying them. But, at that stage, the fee was of less importance to him than the opportunity to test out his rams for a season in flocks across England and observe the results. A large part of his success lay in his precise recording of the feeding and subsequent performance of his animals.

He put great effort into riding around England during the summer, wherever his rams had been sent, evaluating their off-spring and carefully noting down the results. He covered huge distances on horseback at a time when roads were indifferently maintained and travelling off the turnpikes was an uncertain experience. Bakewell must have had a pretty detailed knowl-edge of the English countryside to undertake these excursions at a time before reliable maps. He often travelled alone and unannounced and he would simply observe over the hedge the flocks in which his rams had been at work. This behaviour was characterised by some of his critics as sneaking about and spying on his customers' flocks.

The first ram he let was for 16s for the season in 1760 and he drove it himself to the rendezvous at Leicester Fair. He let two others that year for 17s 6d each. The hiring fees remained modest for the next twenty years or so, until his perseverance overcame the opposition and by 1780 he could boast that he was letting rams for 10 guineas. Then by 1784 he was commonly getting 100 guineas for a ram for the breeding season (about four months). In 1786 he made more than 1,000 guineas from letting twenty rams, one of which was his favourite, 'Twopounder', so called because his shape was said to resemble the barrel of the two-pound cannon. Bakewell let him for a third of the season (probably about a month, or two seventeen-day cycles of ovulation) in the autumn of that year for 400 guineas. And in 1789 he made 3,000 guineas from hiring his rams, when a farm labourer's wage for the half-year was 10 guineas. The *Leicester Journal* reported 'a fact almost incredible' that one of Bakewell's rams had earned 1,200 guineas in hiring fees *in one season*.

He had struggled as a young man against the prejudices of his elders and this kind of success justified the decades of perseverance. When the older farmers and graziers began to hire his improved rams and take his stock seriously, he knew he had made a breakthrough for, as he said, 'when the old birds get into the trap the young ones will follow'. The hiring fees were paid by principal breeders to obtain a stock of rams which they in turn could hire out or sell to commercial farmers and graziers who were producing sheep for the meat trade. Bakewell himself hardly ever sold his rams, except to foreigners for export, because he wanted to keep the breeding within his control, believing it would become diluted and spoiled if it was to fall into hands less skilful than his. He was also shrewd enough not to sell his fat sheep in the livestock markets because he suspected that the butchers would not kill them, but sell them for breeding. So he had them killed at home and sold their meat.

He kept a 'control' sheep in each flock – an animal not of his breeding – that he treated in exactly the same way as his own stock, so that he could compare its performance with his own sheep and thus demonstrate their superiority to his visitors. He also took in farmers' ewes to be served by the rams he had kept at home for his own use. He used 'teaser rams', a practice that he appears to have thought up. These are 'aproned rams of small value' which ran with the flock to detect the ewes that were in heat. The method was to tie a large cloth, or 'apron', around the ram's body in such a position that when he mounted a ewe the cloth prevented penetration. The ewe would then be caught and brought into a small enclosure to be served by the hired ram. This method conserved the stock ram's energy because he was not continually running around trying to detect receptive ewes and it allowed the best rams to serve up to 140 ewes in a season. Bakewell charged between 10 and 60 guineas the score (twenty) for this service. That is between 50 and 300 guineas per ram per season – and these were not his best rams, because most of those would be away on hire.

Bakewell's skill as a showman also contributed to his fame and success, nowhere better exemplified than by the way he ran his annual ram hirings at Dishley at the end of July or the beginning of August. The expectant hirers were assembled in the great barn, which Bakewell had had specially built as a multi-purpose building, two centuries in advance of its time. The rams were brought in singly through one door, paraded round on a stage until all those present had satisfied their curiosity, and then taken out through another door. Once a ram had been hired Bakewell refused to show him again because he said it was human nature to want what one couldn't have, and might cause the hirers to lose interest in the other rams for hire. He also exhibited what he considered to be the worst animals first and progressed to the best. His reason for doing this was, he said, that if he had exhibited the best first 'their great superiority would have made the others [seem] much worse than they really are'.

William Marshall records in 1790 in his survey of the *Rural Economy of the Midland Counties* (vol. I, p. 426) how Bakewell conducted the business of hiring his rams. He invited offers from the assembled buyers, which he either accepted or rejected as he saw fit. Once a price had been agreed there was no writing or legal agreement – everything was trusted to the honour of the parties – and the fee was not to be paid until the ram had had the opportunity to impregnate the agreed number of ewes. About the beginning of September, Bakewell consigned the rams to their hirers, many in places at some distance from Dishley, in carriages 'hung on springs', to ensure they arrived at their destination in the best possible condition. Then, after the ram had done his job, the hirer was expected to return him safely. If a ram died while away on hire, for whatever reason, Bakewell bore the loss. 'The whole system manifested a wonderful degree of confidence and mutual good faith and contributed in an essential degree to the diffusion of the new breed,' wrote Marshall.

Another of Bakewell's ingenious ideas was the 'Dishley Society' or his 'Ram Company' as he called it, which he established in 1783. Bakewell was president until 1795, the year he died. The Society had two main objects: to maintain the purity of the new Dishley breed and to establish a monopoly of breeding rams amongst the members of the Society. If a member refused or neglected to obey the rules the subscription of 10 guineas became forfeit and large penalties were imposed for wilful breach – up to 200 guineas in one case (the equivalent of between £15,000 and £20,000 in current money), which gives a good idea of how seriously its members approached the business of the Society and how much money they must have been making from their sheep breeding.

Some of the rules are: 'no member shall let a Ram on Sundays'; 'secrecy be kept by all members respecting the business of these meetings … and that any member quitting the Society keep secret upon his honor [sic] the transactions before he left it'; and that anyone leaving the room while a meeting was in progress without the permission of the Society 'shall forfeit one shilling for every quarter of an hour he is absent'. The New Dishley Society formed in 1994 continues in Bakewell's memory to promote his revolutionary legacy and encourage research into pioneering agricultural methods.

The French aristocrat François de La Rochefoucauld, who toured England in 1785, paid a visit to Bakewell at Dishley when the great man was over sixty. He thought Bakewell to be 'a man of genius' who early in life had grasped the simple truth about a grazing animal, that the part the butcher sold for the best price was the back – the sirloin and fillet – for roasting, which he said Bakewell described as 'gentleman's meat'. The middle parts were worth less and the lower part 'only fit for the army'. De La Rochefoucauld found it hard to understand how Bakewell had created his improved livestock: 'I don't really understand it, but I believe it, as I believe Religion.' In fact nobody but Bakewell (and his faithful shepherd) really knew how he had done it.

Two years before Bakewell died, one of his great friends, a Mr Paget who was a member of the Dishley Society and a successful breeder of Dishley Leicesters, sold his entire stock of ewes by auction in November 1793. The 130 animals fetched £3,200, an average of £25 16s 11d. That is roughly equivalent, at today's value of money, to about £20,000 for each breeding ewe. The dearest fifteen sheep averaged 52 guineas each – about £40,000. No prices for ewes in the modern sheep world even approach this.

~ 4 ~

THE SWALEDALE

*And he gave it for his opinion ... that whoever could make two ears of corn,
or two blades of grass, to grow upon a spot of ground where only one grew
before, would deserve better of mankind, and do more essential service to his
country, than the whole race of politicians put together.*

Jonathan Swift, *Gulliver's Travels*, II: A Voyage to Brobdingnag

U P IN THE HIGH PENNINES, FOR MOST OF THE YEAR,
the sweeping flat-topped moorland is scourged by
wind and rain and snow; but for a few short months
during the summer, when the land bursts into life and curlews
and lapwings wheel in the high skies, it becomes a magical
landscape, heavy with summer's increase, warm days extending into short, light nights. This is the spine of England that
runs south from the Tyne gap, to Wharfedale on the borders
of Lancashire and Yorkshire, and it is Swaledale country. Small
farms shelter from the storm, their fields parcelled out around
them, bounded by drystone walls and dotted with hundreds
of flagstone-roofed field barns. And beyond the fields, the
open moor.

This high-lying land is good for nothing much except the
extensive grazing of hundreds of thousands of Swaledale

sheep. Swaledales are *the* supreme moorland sheep of the Pennines, so well adapted to the terrain and climate that even their short, loose fleece seems to have taken on the colours of the limestone and the peat as it turns from cream, just after clipping, to weathered ivory later in the season, when it has been darkened by the peaty soil and bleached by the elements.

'They just don't look right anywhere else; if they're bred on different soil they begin to look different – particularly in their wool,' said Matt Mason, whose pedigree flock grazes the moor above Appletreewick in Wharfedale. 'It's hard to know whether they look right because that's what we're used to, or because they really do take their nature from the soil.'

They are very active sheep, with long legs, predominantly white at the front and black on the back, a jet-black face with a white muzzle and white patches around the eyes. Both sexes are horned: the rams' extravagantly curled in up to three spirals, while the ewes' are lighter and more feminine, sweeping back as if they had tucked them behind their ears. They have a remarkable ability to forage and survive on wild moorland and are rather easier to shepherd than most other black-faced, horned hill breeds, because their light, open fleece dries out rapidly after a soaking and makes them less top-heavy and prone to rolling onto their backs and being unable to stand up again. The ewes are tremendously attentive, devoted mothers, capable of producing enough milk to rear a strong lamb from moorland herbage upon which other breeds would struggle to survive.

The Swaledale's characteristics are common to the many breeds of hill sheep in Britain, albeit each is adapted to its particular terrain. In common with all hill breeds, although they do provide meat and wool from poor land, exceeding anything that any other creature or activity could produce for human benefit, they are slow to mature and do not grow very big. But their value has been transformed over the last century or so, by what has come to be called the *sheep pyramid*. This is our unique national system of cross-breeding which uses the innate genetic potential of hill sheep to extend their productive lives and produce more lamb for the national larder.

Diagrammatically this could perhaps be represented by *two* flat-topped pyramids – one inverted, with its top resting on the one below – in a shape like an egg-timer. At the highest level are the millions of pure-bred hill breeds, and in the middle the half-bred ewes from the first cross with Longwool crossing breeds, and the wide lowest band represents the millions of meat lambs born to those half-bred ewes that have been crossed with a meat-producing Down ram (the 'terminal sire'). The effect of this arrangement is that most of the lamb produced for the table in Britain is descended from one of our pure breeds of hill sheep, which form the genetic reservoir upon which the productivity of the modern British sheep industry depends.

The system is based on a genetic phenomenon called *hybrid vigour*, or *heterosis*. This is the effect of the first crossing of two pure breeds which combines and enhances the best characteristics of each parent and subordinates their less desirable ones in a hybrid that is superior to both parents. Every time the cross is done the progeny always have the same characteristics, so long as their parents have been bred pure for long enough. Most enhanced agricultural production across the world depends on hybridisation. Nearly all our eggs and poultry come from hybrids, as do most of our commercially grown vegetables. For example, the grapefruit is a hybrid of a Jamaican sweet orange and a south-east Asian citrus fruit called a pomelo, or pompelmous (French *pamplemousse*); and peppermint is a naturally occurring sterile hybrid of spearmint and water mint. This first cross of two pure breeds is an *F1 hybrid* – 'F' meaning filial and '1' the first generation.

One of my farming neighbours used to refer to the

beneficial effect of 'highbred vigour', which inadvertently catches the essence of its invigorating effect. When it works, the first cross always inherits predictable and recognisable characteristics. So, crossing a Swaledale ewe with the Bluefaced Leicester ram, for example, always results in the astonishingly successful hybrid called the Mule, from the Latin *mulus*, via the Old French *mule*, meaning hybrid. It is not clear why the Bluefaced Leicester × Swaledale is the only hybrid to have been given the name Mule.

Not every first cross produces better offspring than its parents. Crossing a Swaledale ram with a Bluefaced Leicester ewe, for example, gives a decidedly unsatisfactory result. And although many first-generation plant hybrids are sterile, animals are often capable of breeding a second and subsequent generations from the F_1 offspring (the F_2 generation and so on), but neither does this give a predictable result, nor preserve the vigour of the first cross. Sometimes the less desirable characteristics of the pure breeds can resurge in the third and later generations, when an unwelcome trait is said to have 'skipped a generation', or an individual is described as a 'throwback'.

The important point is that crossing to create predictably uniform hybrids requires a continuous supply of parent breeds that have been bred pure for many generations on both sides. And that is why the pure hill breeds are such a valuable resource and why distinct sheep breeds emerged about 150 years ago. Breeders strove to fix the characteristics that had emerged as a result of long association with a particular locale, because they needed to create a more regular uniform type than when the emphasis had been simply on wool.

The British Isles are ideal for this kind of integrated system because we have a larger climatic range in a smaller area from north to south, and sea level to mountain-top, than nearly any other country in the world, including some of the best land in Europe with a good deal of the worst. The genius of the sheep pyramid is that it brings all our nation's land into a productive whole: the tracts of semi-tundra on our mountains (with a climate similar to the Arctic Circle's), the uplands and the lush pastures of the lowlands. The railways in the nineteenth and the motorways in the last century also played their part by shortening the distances and making it possible to move livestock, in better condition, at a fraction of the cost of droving.

Imagine you are the owner of a flock of 1,000 pure-bred Swaledale ewes which have become acclimatised to your wild, windswept farm on the high Pennine moors over innumerable generations. Each ewe rears, on average, one lamb each year – rather fewer than the 125–30 per cent that would be usual, but it makes the maths easier – half and half males and females. (In fact slightly more males than females are born – the ratio is about 53:47). Almost all your male lambs will be castrated (made into wethers) soon after birth and be sold as the grass stops growing in the autumn, as *stores*, i.e. animals that are not yet ready for the butcher (*finished*). They will be bought by farmers with better, lower-lying land, for further feeding.

Apart from the wethers there will be 500 ewe lambs, and of course you will still have your original 1,000 breeding ewes – less the 3 per cent that have died during the previous year from various causes. If you were to keep all the ewe lambs your flock would double in size in two years. So what do you do? The answer depends on a judicious balance of youth with quality, and the success of your enterprise depends on how well you weigh one against the other.

The objective is to keep the flock as young as possible, because young sheep are healthier and more resilient, and always to strive to improve the quality of your flock. You would not want to sell off strong, well-bred older ewes if their teeth and feet were still good, so you might keep them for a year or two longer than others. Whatever your policy, you would have to sell about 470 female sheep every year just to maintain the size of your flock, and most of these would be the older ewes. So let us assume that you keep the best 370 ewe lambs and sell the poorest 130. You will also have to sell 370 of the older ewes to make way for the ewe lambs you have kept to replace them.

In August, when the lambs are weaned from their mothers, you will decide which ewes and lambs to keep and which to sell. To do this you will gather up the flock, separate the ewes from the lambs, and inspect each animal. The ewes are then dipped and taken back onto the moor without their lambs. This allows them to get over the separation and their milk to dry up. Once they are weaned the lambs become 'hoggs', until their first clipping next year, when they become 'shearlings'.

The lambs stay in the fields for a few days, on better grazing – generally the re-growth after the hay has been taken – until they, too, are accustomed to the separation. Once settled, they are sorted into four categories: the wether lambs for sale, the ewe lambs to keep for breeding and the poorer ewe lambs for sale. You might also have kept a few ram lambs for sale to other farmers for breeding. After two or three weeks, the ewe lambs you are going to keep will be sent back to the moor until tupping time, and those for sale will be kept in the fields until sale time. The ram lambs for sale for breeding will be preened and pampered like babies.

Once the ewes have had a few weeks back on the moor without their lambs, and they are in the best condition to face the winter, they are gathered again to draw out (thus *draft* ewes) those to take to the autumn draft ewe sale. These sales have been held all over the country for many centuries, usually in towns on the edges of the uplands. Nowadays they are simply sheep sales, but in times past they were fairs, great social events where people got together, transacted business, renewed their acquaintance and amused themselves. The autumn sales are still great events in the flockmaster's calendar, when most of his income comes in, marking the end of the breeding year before the cycle begins again with the rams going in with the ewes. For pastoral people the end of October is the end of the year, when their flocks and herds are at their best and when the shepherd exposes his year's work to the criticism of his neighbours and the hard evaluation of buyers from all over the country.

Most of the draft ewes will be no more than five years old and will have had three or four lambs. Hill sheep of this age still have a lot of breeding life left in them. If they are off hard hills, many of the ewes will grow bigger when they get onto better land, where the climate is kinder and the grazing more nutritious. Before the development of the sheep pyramid, they would have been sold at modest prices for slaughter. But the buyers at the sheep sales know that the appearance of these ewes belies hidden qualities. They are about to begin the second stage of their lives, when they will not only improve in condition, but the more nutritious grazing will stimulate them to shed more eggs at ovulation, causing them to conceive twins and, quite often, triplets rather than the single lambs they bore on the hill.

In early November the draft Swaledale ewes you have sold, to go onto better land, will be introduced to a Longwool crossing ram, usually a Bluefaced Leicester. If they are well-managed they will conceive, on average, and give birth to, two Mule lambs each. The male Mule lambs are all destined for the butcher. But the females, which are the object of the breeding exercise, will be pampered, preened and dipped, have their faces and legs clipped and washed, and be sold for high prices

63

at the autumn sales. Lowland farmers with even better land can offer them still better feeding, either on the by-products of arable production – sugar beet residue, for example – or intensively grazed grass. These females are valuable for their inherited qualities: from their Swaledale mother they get their hardiness, and milking and mothering ability; and from their father, fecundity, size, conformation (shape of carcase) and their lustrous wool. That is the first cross in the sheep pyramid.

For the second cross the Mule ewe lambs are mated with a specialised meat-producing breed such as a Suffolk or a Texel, or one of the new French importations like the Charollais (which inherited many of its qualities from a cross with the New Leicester). They will each produce two superb fast-growing butchers' lambs for the table for five, six or more years. This third and final stage of the sheep pyramid is accurately, but rather indelicately, called the 'terminal' stage.

This is how the classic sheep pyramid works with Swaledale sheep. All the other pure-bred mountain and hill breeds (with certain exceptions) are similarly managed and collectively they form the genetic source of pure breeding stock upon which the whole cross-breeding system depends. The extra profits it generates, derived from increased productivity, flow up and down the pyramid. The hill farmer with a pure-bred flock gets a much better price for his old ewes than he would if they merely went for slaughter, and their productive lives are extended (often doubled). The first-cross hybrid ewe is a much more productive animal than pure-bred hill breeds or Longwools would be on their own, and the farmer selling butchers' lambs gets more lambs of better quality from his breeding sheep that live longer than pure-bred animals. There are other, more subtle benefits; for example, the annual sale of his draft ewes imposes a sort of discipline on the hill farmer to keep his flock young, through an autumn clearing-out of the older ewes. Rather like the free circulation of money being necessary for the health of an economy, the spreading of nature's increase generates production and keeps its bounty flowing.

In England and Wales there is a much wider array of hill breeds than in Scotland, partly because they have been developed over a much longer time, but also because there is a wider range of topography and soil types and a more variable climate than in Scotland. Although the different hill breeds are equally capable of turning the wild places into profit, some are better suited to hybridisation than others. Much depends on whether a suitable Longwool crossing breed has been found to match them and carry through their best characteristics into their cross-bred daughters.

Upper Wharfedale, near Skipton, is at the south-western edge of the Swaledale's territory. It has been sheep country for thousands of years – even Skipton, *skip-tun*, means sheep farm – which it was long before the huge monastic sheep-walks of the Augustinian monastery of Bolton Abbey carried thousands of sheep, producing wool for the medieval cloth trade. Since the Dissolution, Bolton Abbey and its beautiful 30,000-acre estate has been owned by the Dukes of Devonshire, whose tenants, until recently, kept flocks of Swaledales every bit as numerous as anything the monastic houses ever kept. But in recent years, for the first time in many centuries, about a third of the sheep have been removed from the moors to satisfy the EU policy of reducing the flocks of sheep on the uplands, causing large areas of moorland to be overgrown with knee-deep herbage.

This part of Wharfedale is delightful: limestone houses with stone-flagged roofs and mullioned windows; everywhere the wide skies and sweeping moors above fields zigzagged by drystone walls and studded with field-barns. The Mason family has been farming at Appletreewick in Wharfedale since 1956, when Matt's father bought the 380-acre farm (plus grazing rights on the open moorland) for £8,000. For months afterwards his father lay awake at night worrying how he was going to pay back the money. There have been Masons in nearby Dentdale since 1492. The first Mason is supposed to have been a bowman in Duke William's army at the Battle of Hastings, and it is not hard to believe. Matt is a powerful, chestnut-bearded, forthright, independent man who has kept pedigree Swaledale sheep for over forty years. At one time he milked cows as well, but he gave that up a few years ago, 'getting rid of that dairy herd was like being let out of prison', he said.

Like many other farmers in the northern counties Matt lost all his stock, except ninety sheep that were away from the farm, in the foot-and-mouth cull in 2001, which he thinks was a kind of watershed for traditional hill sheep farming. Many of the best sheep, descendants of flocks that had been in the hills for centuries, were lost. Matt was particularly hard hit because he had been selling very high-quality sheep of ancient pedigree. His rams regularly made £15,000 and his draft ewes up to £350 each – over four times the average. He has found it hard to breed back into the same quality of stock, let alone accustom them to his hill land. Farmers do not usually sell their best breeding stock, so it is hard to find high-quality sheep to replace them. It takes a long time to accustom sheep to the terrain by natural adaptation and domestic selection. Who now has the time, or the money, or the faith in the future, to acclimatise new sheep to the hills?

Swaledale breeders were some of the first to recognise the value of organising themselves into a breed society and keeping a flock book to record the pedigrees of breeding sheep. The Swaledale Sheep Breeders' Association was founded in 1919 by a group of breeders living within a seven-mile radius of Tan Hill pub, at 1,730 feet above sea level, the highest public house in England, where the borders of Co. Durham, Yorkshire and Cumbria meet.

In the preface to the first Flock Book, the Swaledale Breed Association made an admirable attempt to describe the characteristics of a good Swaledale. But it is hard to know whether it is describing what the members of the Association were hoping to achieve by domestic selection or whether they were simply describing what Swaledales would look like if they were acclimatised to the moors through enough generations. Alan Alderson, the current chairman, explains rather enigmatically that a good sheep 'just looks right'. Then when pressed he surprised me with a more lyrical reply:

'If you were comparing the texture of their wool and the texture of the grasses they're eating – if you were looking at it in black and white and you could put your hands into either, you would feel comfortable with the fleece *or* the grass, as if they were one and the same thing.'

There are some privileged Swaledales that never have to endure the rigours of life in the hills. One such is 'Private Derby', the pampered ram that is the regimental mascot of the Mercian Regiment. The first Private Derby (not a Swaledale) was captured by the 95th Derbyshire Regiment of Foot in April 1858 at the siege of Kotah during the Indian Mutiny campaign of 1857–8. This ram marched over 3,000 miles with his regiment, fought thirty-three undefeated battles against other rams, and was awarded a campaign medal. Since then the regiment has had an unbroken succession of Private Derbys, the latest being Derby XXIX, presented to the Mercian Regiment in 2009 by the Duke of Devonshire, who since 1912 has presented to the regiment every Private Derby from his Chatsworth Park flock.

Private Derby has a Ram Major and a Ram Orderly, appointed from the Drums Platoon, who are responsible for his welfare. He has a regimental number and draws his rations like any soldier. When on parade the Ram Major and the Ram Orderly lead him, one on either side, by two white ornamental ropes, attached to a leather collar. He wears a scarlet coat, with buff and gold facings, emblazoned with the regiment's battle honours, which include his Indian Mutiny medal, his General Service Medal 1962 (with Clasp Northern Ireland, where he has been stationed over the years), a silver plate on his forehead embossed with the Regimental Cap Badge, and a pair of silver protectors on the tips of his horns.

🐏 5 🐏

THE CHEVIOT

In all the different districts of the kingdom we find various breeds of sheep beautifully adapted to the locality which they occupy. No one knows their origin; they are indigenous to the soil, climate, and pasturage, the locality on which they graze; they seem to have been formed for it and by it.

William Youatt, *Sheep: Their Breeds, Management and Diseases,* 1837, p. 312

I T IS IN SCOTLAND THAT THE SHEEP PYRAMID BECAME refined, almost to the point of perfection, and was developed on a huge scale. To understand the way this played out it is necessary to see Scotland as a tale of two very different halves: the land south of the central lowlands and the Highlands. Although a type that is the Cheviot – white-faced and fine-woolled – had been grazing the grassy hills of the Borders and the Southern Uplands for many centuries, in the rest of Scotland north of the central lowlands the indigenous sheep kept by the crofters had not changed much since the Iron Age. They were not numerous, because the predominant domestic livestock was cattle, but they had a valuable place in the crofting economy for their exceptionally fine wool, and would have been found, with regional variations, all over Scotland.

In 1746, after the Battle of Culloden, when the clan system was swept away, the crofters and their *seana chaorich cheaga* (little old sheep) rapidly followed. Huge tracts of the Highlands and Islands were emptied of people. And rather in the same way that the American prairies, once denuded of their native inhabitants, were exploited as the bread-basket of the society that replaced them, the empty land in Scotland was rapidly colonised by the *caoirich mhora* (the big English sheep) that fed and clothed the growing towns of the British industrial revolution. From the second half of the eighteenth century there flowed from the hills of Scotland a bonanza of sheep and wool that lasted almost two centuries, and has declined only in recent decades.

The two breeds of big English sheep that have vied for supremacy in the Highlands of Scotland are the Scotch Blackface and the Cheviot. They are the yin and yang of Scottish hill sheep, and descend from entirely different ancestors. The Cheviot is the northerly representative of a dun-faced type kept by the Celtic peoples long before the Romans came. It is native to the Scottish Borders country, very hardy, with probably the best shortwoolled fleece of all the hill breeds. The ewes are hornless (*polled*) with erect white ears, white faces and legs and black noses and feet. Even as early as 1795 one observer wrote of the Cheviot, 'the same kind of polled sheep have fed in this district from time immemorial; nor does anybody alledge [sic] that they were ever natives of any other region.'

In the south of Scotland the Cheviot was distinguished from the Blackface by being called the 'long sheep' and the 'short sheep'. There is an instructive anecdote in *The Domestic Manners of Sir Walter Scott* where James Hogg recounts a meeting between Scott and Mr Walter Brydon, a renowned breeder of Cheviot sheep, on one of Scott's excursions to the Ettrick Forest to gather stories and ballads:

The original black-faced forest breed being always denominated the *short sheep* and the Cheviot breed, the *long sheep*. The disputes at that time ran very high about the practicable profits of each. Mr Scott ... felt himself rather bored with the everlasting question of the long and short sheep. So, at length, putting on his most serious calculating face, he turned to Mr Walter Brydon and said, 'I am rather at a loss regarding the merits of this *very* important question. How long must a sheep actually measure to come under the denomination of a *long sheep?*' Mr Brydon, who, in the simplicity of his heart, neither perceived the quiz nor the reproof, fell to answer with great sincerity, 'It's the woo' sir; it's the woo' that maks the difference, the lang sheep hae the short woo' an' the short sheep hae the lang thing, an' these are just the kind o' names we gie them, ye see.'

The Cheviots' grazing over millennia across the Southern Uplands probably destroyed the Wildwood – the ancient natural woodland that had clothed Western Europe since the last ice retreated about 15,000 years ago – and prevented its regeneration by eating the seedlings that would have replaced the trees as they matured and died.

We know from Roman writers that the pastoral Celts had considerable flocks in the Borders and practised a short-distance form of transhumance, a system which endured almost into living memory. During the winter the sheep grazed the low-lying fields in the 'winter town'. Then in spring they were taken up to the 'spring town' in the hills to crop the high-lying grazings. This annual migration began on the same day every spring when the young men took up the cattle. The young women followed a few weeks later with the ewes, which they milked in 'yowe buchts'– low, turf-walled enclosures – to make hard cheese ('white meat') for the winter. In parts of the

Borders cheese from Cheviot ewes' milk was made well into the nineteenth century.

After the Conquest, when religious houses acquired large estates in the Borders, they encountered an existing pastoral economy of ancient standing, and often found themselves in competition for grazing rights. The Cistercians began by keeping sheep to produce undyed wool for their clothing, particularly their habits, hence their sobriquet 'the white monks'. Then, almost incidentally, their prowess at sheep keeping provided them with a large income from the surplus of wool that far exceeded their own needs. Francesco Pegolotti recorded in his handbook that wool from the Cistercian house at Mirososso (Melrose) was of the highest quality, better than the other abbeys from which he bought wool, at Ghelzo (Kelso), Guldingamo (Coldingham), or Gridegorda (Jedburgh).

These monastic sheep flocks grew to be considerable. The Cistercian houses alone produced 20 per cent of Scottish wool. Melrose had 12,000 sheep and Kelso Abbey's Rent Rolls for 1290–1300 record that they had between 7,000 and 8,000 sheep on the open hills around the Bowmont valley. The records suggest that there were about a quarter of a million sheep in the Borders at that time, with wool of such quality that European merchants were prepared to forward-buy it a number of years ahead. In some seasons when the wool-clip was deficient, or failed to make the grade, the abbeys would have to buy wool in the domestic market to honour their contracts. Their wool was shipped out down the Tweed to Berwick, the main Scottish seaport at the time – probably in shallow-drafted coracles which could negotiate the river – from where it went by sea to the manufactories of Flanders and Italy.

Wool became so valuable in the early Middle Ages that every available acre in the Borders (and in many other places in Britain) was grazed by sheep, hundreds of thousands of them, relentlessly eating their way across the wide hills. This increase in the sheep population may have coincided with a period of warmer weather, the 'medieval warm period' that is supposed to have lasted from about 1000 AD until the beginning of the fourteenth century.

But all this came to an abrupt end with the Battle of Bannockburn in 1314 when the Borders descended into 300 years of lawlessness that destroyed civil society and all hope of a settled agriculture. It was not until 1603, when James I acceded to the throne of England, that peace was restored. But even then it took the Act of Settlement in 1707 before farmers could once again keep sheep on the Borders hills without the constant fear of having them stolen, their families murdered or their houses and lands laid waste. The peace allowed the Borders flocks to regenerate and set in train a zeal for land improvement and livestock breeding that endures to this day.

It also saw an extraordinary flowering of Borders woollen manufacture based almost entirely on the fleeces from Cheviot sheep. During the height of its medieval production much Borders wool had been exported. But this trade declined during the three centuries of warfare, with local production going to support a modest cottage industry making rustic cloth for local consumption. The undyed wool was worked into a traditional pattern of checks in white and grey or brown, called the 'Shepherd's Plaid'. This had been the traditional dress of countrymen all over Scotland for many centuries. James Hogg and Walter Scott wore this type of plaid, a sample of which is displayed in a glass case at the Duke of Buccleuch's Bowhill House, near Selkirk. It was similar to the pattern of cloth described by Tacitus as the dress of the Celtic tribes the Romans encountered in the Borders.

The wool was sorted into hanks of natural shades and woven into the weft and warp of the cloth (rather like a dog's-tooth pattern) in a design that was known as 'checks', or 'Galashiels Grey' in the Borders. A similar type of cloth, although in a plainer style, called 'Hodden Grey', was made in Cumberland from the wool of Herdwick sheep. Each region had its distinctive traditional design that would have advertised its wearer's origins just as surely as would his accent.

Then, in the eighteenth century, just as the dominance of wool was beginning to yield to meat in England, three Scottish inventions transformed the cottage weaving into a world-renowned woollen industry. These were the flying shuttle, allowing much wider fabrics to be woven, the 'Teazing Willy' (a machine that made wool easier to spin) and the water-driven power-loom. These caused a great building of water-powered woollen mills at Galashiels, alongside the Gala Water at its confluence with the Tweed, and set in train the huge international trade in tartan and tweed.

Tweed owes its phenomenal success to two Borders men: Archibald Craig and James Locke. Craig was already a moderately successful cloth merchant in London when he saw the potential in the metropolitan fashion for rustic checked cloth. The story goes that he received a stained, sub-standard sample of black and white checked cloth from one of his artisan manufacturers in the Borders. Rather than send it back, he dyed it brown to disguise the staining and turned it into a brown and black check. He then tried green dye, to make a green and black check, and blue for blue and black. These proved so popular that he had the weavers make cloths in various sizes of check and style experimenting with new patterns and colours.

But the real marketing genius was James Locke, an enterprising Scottish clothier, who traded from premises in Regent Street. The fashion at the time was for jackets and trousers made up from different patterns of cloth; morning dress is a survival of this. But when the Prince of Wales appeared in public in a jacket and trousers that 'suited' one another – made from cloth of the same pattern – a new sartorial fashion was set and Borders manufacturers satisfied it by producing lengths of

'tweel' (twill) 'suiting' in huge quantities. In 1847 Locke is supposed to have received a consignment of 'tweel' from Messrs.
William Watson & Son of Hawick. On the note attached to the
parcel the word tweel read 'tweed' because an ink blot at the
base of the 'l' made it look like a 'd'. Locke wrote back asking
for some more of their 'tweed' and thus the brand was born.
Sir Walter Scott had already done much to make the Borders
and the Tweed valley famous all over the world, but making
the name of the cloth synonymous with the river was a brilliant stroke and transformed tweed into a unique worldwide
brand and ensured it a permanent place in the wardrobe of
anyone who aspired to gentility. Tweed has never since lost its
worldwide appeal

Tartan has an even more fanciful origin. In 1815 the secretary of the Highland Society (a club of expatriate Scotsmen)
wrote to every clan chief he could find, asking for a swatch of
his clan 'tartan' so that it could be 'registered' and archived.
It made no difference that most of the chiefs had never heard
of tartan, and were baffled by the request, because there were
plenty of manufacturers willing to remember their pattern
for them. The enterprising Sobieski-Stewart brothers, John
and Charles-Edward, half-Polish, half-Scottish adventurers,
who claimed tenuous descent from Bonnie Prince Charlie,
compiled a catalogue of all the 'ancient clan tartans' in their
wholly invented book, *Vestiarum Scoticum*. This appealed pow-
erfully to Scottish sentimental nationalism, both at home and
abroad, and ever since tartan has been almost obligatory, at
social events, for nearly all classes of Scottish society.

George IV sealed tartan's appeal in 1822 by making the first
state visit of a reigning monarch to Scotland since Charles II in
1651. Sir Walter Scott was deputed to organise the whole affair
and is supposed to have persuaded the King to deck himself
out in Royal Stewart tartan. When his Majesty's portly figure
proceeded down the gangplank at Leith docks it was swathed

in a tartan plaid over a kilt, topped off with a matching bonnet. This set a trend that subsequent monarchs have followed, to considerable effect, as a unifying symbol. Queen Victoria and Prince Albert were enthusiastic wearers of tartan, as have been many subsequent members of the Royal Family.

Sir Walter Scott himself would never wear tartan, particularly disliking the kilt. His habitual dress was 'breeks' made from the shepherd's plaid. His son-in-law and biographer, J. G. Lockhart, shared Scott's distaste for the kilt; he thought his father-in-law had created a caricature of the Scots as 'a nation of Highlanders, and the bagpipe and the tartan are the order of the day'.

As this wool boom went almost hand in hand with the growth in demand for mutton, Borders flockmasters contrived to respond to both. The Cheviot was so wonderfully well adapted to its native hills that it was a relatively easy task to modify it to satisfy the demand for meat by careful cross-breeding. They sought out 'improving tups' to put to their Cheviot ewes, to improve the breed's carcase and bring it to earlier maturity and, in so doing, they achieved something that had eluded many other improvers, namely they managed not to mar the wool too much. In the 1750s three prominent Borders breeders, John Edmistoun, Charles Ker and James Robson, travelled to Lincolnshire and paid high prices for suitable rams. These were probably of the medieval Lincoln type, similar to those that Bakewell crossed with the native Leicester, to create his New Leicester. These farsighted breeders (particularly James Robson) are credited with being the creators of the modern Cheviot. It was probably at this time that the Cheviot lost its mottled tan face and acquired a pure white face and legs. Crucially, the improvements did not sacrifice the invaluable traits of the hill ewe: thriftiness, hardiness, devoted mothering and an almost preternatural ability to predict the approach of a storm and move to shelter from it.

Hill flocks in Scotland and the Borders are divided into hirsels. These contain an average of about 400 ewes, depending on the terrain. Hirsel derives from the Old Norse, *hirtha*, to herd or tend, and refers to that quantity of sheep a shepherd can comfortably care for on the hill. It has also come to be synonymous with the area of land upon which the sheep live. The daily management of a hirsel includes raking the flock in the morning and again in the evening. *Raking* comes from the Old Norse *raka*, to drive, which describes putting the dogs behind the sheep to drive them from the bottom of the hill to the top in the morning and leaving them to graze their way back down the hill during the day. In the evening the shepherd would then bring the flock together, gathering in any stragglers. Raking has a number of benefits: it encourages the flock to graze evenly across the whole hirsel and make better use of the land than the sheep would if left to their own devices, and discourages them from concentrating on the areas where they find the sweetest grazing; it also keeps them active, which is important during pregnancy, and allows the shepherd to see every one of his sheep twice a day, and by moving them and counting them the shepherd can immediately see if any is ill or lame or missing. Now that shepherds are expected to manage much larger flocks, they often do not have time for the intensive shepherding that would have been possible when each shepherd only managed his own hirsel.

As sheep proliferated and became more valuable during the eighteenth century, it became necessary to find some way of protecting them from the dreadful weather that the Scottish hills can throw up. The answer in Scotland, but rarely found anywhere else, was to build stells to accommodate and shelter the sheep on each hirsel. *Stell* is the Scottish way of pronouncing stall (as in cow stall) and means a fixed place. These usually circular structures enclosed a few hundred square yards of land with stone walls five or six feet high, sometimes protected by a planting of trees, where the sheep and the shepherd could find protection from storms. In some places the stell was built in a cruciform shape, without outer walls, so that the flock could shelter out of the storm wherever it came from. Stells saved many thousands of sheep during harsh winters. Sheep are not forced into them, but they resort to them when driven off their grazings by a storm. They can be fed within the walls when the land is too deeply covered with snow for them to reach the herbage below. The walls were built so that driving snow in a blizzard would tend to drift against the windward wall, then blow right over the top rather than accumulate inside the walls. Even if it did drift over the flock sheltering inside, the shepherd would know where to find them to dig them out if he had to. The capital cost of erecting stells was small compared with the value of the flock, and was recovered ten times over by the numbers of sheep saved in a bad winter.

The centre of the Cheviot's ancestral territory is the wide grassy hills around the headwaters of the River Coquet, where the land makes better Cheviot sheep than anywhere else. This is thought to be because the soil overlies an intrusion of red porphyry rock, which can be seen in the bed of the Coquet, hence its meaning, 'red river', the *coch* prefix coming from the Brythonic Celtic for red, as in cochineal. Porphyry is rich in dissolved minerals necessary for the health of grazing animals, and soils overlying it contain a mixture of them, particularly copper, cobalt, molybdenum and selenium.

The Cheviot is usually polled in both sexes, although in the past, rams would have been horned; some still display vestigial horns, despite breeders' efforts over generations to eliminate them. The Cheviot Breed Society allows that horns are 'permissible but not desirable', recognising the difficulty of extirpating them without losing the breed's integrity. Every Cheviot must have white legs, white face and a black nose with 'a bold dark eye'. A pink nose indicates softness, as do pink cleats, which predict foot troubles.

As with all cloven-hoofed herbivores, sheep have two hooves on each foot upon which they walk on tip-toe (with a cleat (cleft) between the two). The hooves grow like finger-nails, are hard at the extremity and need to be worn down by coming into contact with an abrasive surface. In the skin of each cleat there is a gland that exudes a scent which is passed onto the ground as the sheep walks. Each member of the flock recognises the scent given off by the glands of other members of the same flock and, in a way not understood, this encourages their flocking instinct. Some breeds have a better-developed flocking instinct than others. Primitive sheep like the Soay and Jacob, for example, tend to scatter when threatened, whereas more highly bred types flock together more tightly.

Another attribute esteemed by breeders is oval leg bones ('flat', as they describe them). Oval is stronger than round and an indication of superior milking ability. Black dots or 'ticks' on the skin, particularly on the face, are valued as a 'stylish' indication of hardiness. These are examples of Cuvier's Law of Correlation, which connects the presence of one obvious characteristic with some secondary, often hidden, trait: for example, white cats with blue eyes are almost invariably deaf, and Herdwick lambs born jet black with pure white ears will be stronger and better sheep than their pure black-eared con-temporaries. It may also explain why breeding off the horns can destroy the integrity of a breed.

In essence, it means that one obvious trait in an animal indicates that others will be present in that animal's makeup as part of a balanced and integrated whole. Bakewell knew that a sheep's carcase could not be changed without its wool deteriorating, because to improve the one necessarily causes the other to deteriorate. At first sight this seems to run against Darwin's theory which suggests that organisms are infinitely malleable and that by domestic selection (i.e. selective breeding) one characteristic can be altered independently of all the others which will remain unaffected. But Darwin did not reject Cuvier's Law; on the contrary he recognised 'that when one part varies, and the variations are accumulated through natural selection, other modifications, often of the most unexpected nature, will ensue'.

Lug marks are the traditional way that sheep owners marked their sheep. They belong to the farm and not to the owner, who is merely the guardian for the time being of the flock that lives there. That is one reason why at the autumn sales of draft ewes the name of the farm is given equal prominence in the catalogue to the name of the farmer. They have to use ear tags as well now, because the EU has made it compulsory, but many farmers do not like them because they tend to make the lambs' ears fester and then the tags fall out. The plastic tags required by the EU earmarking regulations are unsuited to the erect ears of Cheviots because piercing often damages the cartilage, causing the ear to flop over, which destroys the distinctive 'cock o' the lug' that goes with the 'glint o' the eye' that is so characteristic of the breed. When a shepherd looks at his sheep, it is always defects that catch his eye, because they indicate illness or distress, and a damaged, flopped-over ear is one of the first things he will notice, and keep on noticing every time he inspects his flock.

Lug means *ear* in the north and it is derived from the Old Norse *lög*, meaning law. So the lawful mark has come to mean the place where the mark is made. It is done more to prove ownership to a stranger, than to the owner, because a shepherd knows his sheep and recognises them on sight, although he might occasionally have to check a lug mark, just to resolve any doubt. Lug marks are made in various combinations of slits, notches and holes, each with its own description. A 'bit' is a notch, and a 'fold bit' is exactly what it says: the ear is folded like a piece of paper and the corner cut off to make a 'V' shaped notch. 'Key bitted' is a square notch and 'bitted' is a notch which looks as if a bite has been taken out. The ear can be 'halved' – half the end cut off, 'forked' – a 'V' shape cut out of the end, 'cropped' – the end cut off entirely, 'slit' when a cut is made half-way down the length of the ear, or 'ritted' – when a sliver is taken out down the length from the end to divide it into two halves, or 'twice ritted', which is pretty drastic. It looks barbaric, but it can't be much more painful than the piercings and tattoos in fashion today.

The lambs' ears are cut at dipping time, after lambing, with a pair of sheep shears or a special pair of pliers. If the lord of the manor (the owner of the soil of common land) kept a flock on a common, he was the only grazier whose sheep did not have to be lug marked, so that any animal with clean ears belonged to the lord. And any animal with both ears cropped, so that all distinguishing marks had been removed, was forfeit to the lord; the idea being to make it pointless for sheep steal-ers to cut off a sheep's ears because whoever it really belonged to, it certainly could not belong to the man who claimed it.

Hardiness, resilience and the sheer will to endure are the all-important attributes if sheep are to survive the winters on the bleak hills of Tim Elliot's farm at Hindhope, which can be blanketed with snow for two or three months of the year, often to a considerable depth, during which the sheep are expected to scrape through to reach the roughage below. If there is plenty of accumulated tussocky grass from the previ-ous summer and the sheep can fill their bellies, then they will usually survive the severest winters.

On traditionally managed hill sheep farms such as at Hindhope, where the flock lives almost entirely off what grows naturally, the number of sheep is balanced by long experience with the natural growth of grass. The skill of the flockmaster is to ensure that by the end of the summer there is enough roughage remaining to sustain the flock through the winter. Then, just before the grass starts to grow again, the sheep should have eaten off last year's accumulated roughage and the land be ready for the new growth. May is the only month of the year when hill grazings ought to be bare. This is because most of the natural sedges and grasses grow from the middle of May to the middle of October and because there is little growth outside that period the winter's store of roughage should have accumulated by the time growth ceases and should last until the spring.

Lambing is carefully timed to make best use of the first and most nutritious spring growth. Most breeds (apart from the Dorset and a few with Mediterranean ancestry) come into season once every seventeen days or so, for about three months in the autumn. Hill sheep have a shorter breeding season than lowland breeds, and for Cheviots the window of breeding opportunity is from about the end of September until the end of December. Gestation is roughly 147 days, or five months. At places like Hindhope, which are nearly all open hill land, there is no room for all the ewes in the fields at the same time. So they stay on the hill for lambing and the shepherd goes round them at least twice a day. It is a considerable job, even for three men, to do the rounds of 1,400 lambing ewes that are ranged across 3,000 acres of rough, open hill land.

The rams are not allowed in with the ewes until 20 November to ensure that the first lambs do not arrive before 18 April, so that when the first shoots of new grass start to grow about the beginning of May, the first lambs, at two weeks old, are just capable of making the best use of the flush of milk that it stimulates in their mothers. In theory the whole flock could lamb during the seventeen days following 18 April, but in practice this never happens, which is a mercy for the shepherd, who would otherwise have to deal with all the ewes lambing in two and a half weeks. On the other hand if lambing extends over too long a period it prevents the flock being managed as a whole, and late lambs miss the advantages of an early start in the year, a longer grazing season and more sunshine. In practice most lambs are born within about six weeks (two cycles of ovulation) and nearly all the ewes will have lambed by the end of May.

Most ewes lamb without any human interference. In fact the more they are bothered the more trouble they tend to give. A ewe, left alone on the hill, when her time comes, will separate herself from the flock, find a sheltered place – a hollow, or the lee of a rock – make a rudimentary scrape in the ground – an atavistic concession to her wild ancestry – lie down and give birth. Hill ewes normally do it quickly and easily because they have inherited that quality through natural and domestic selection. Those that didn't lamb easily either died in lambing or their lamb did. Almost immediately after giving birth the ewe is on her feet and licking her lamb clean of its birth fluids. This licking cleans and stimulates the lamb to struggle to its feet to take its first milk. All this will usually be accomplished within less than half an hour. The lamb's suckling causes its mother's womb to contract and expel the afterbirth, which she will usually eat – if the foxes, badgers, crows or buzzards don't get to it first.

The hill shepherd's role is rather like a policeman on the beat – keeping an eye on everything that's going on. Even though only a small proportion – 10–15 per cent – of hill ewes ever need help, when they do they are harder to deal with because they are out on open land, often a long way from any sheep pens or buildings. You can usually tell from her behaviour that there's something wrong with a ewe, but you don't know exactly what it is until you catch and examine her. That is where your dog comes in. A good dog on an open hill can, by force of personality, prevent a single ewe from running off, blocking her escape long enough for the shepherd to catch her with his crook. There are some strong dogs that can catch a ewe by gripping the wool at the neck and keeping hold until the shepherd can take over.

A ewe usually has trouble lambing because the lamb isn't positioned correctly in the birth canal. The correct presentation is to have the lamb's body elongated, so that it is as narrow as possible, with front legs together, like a diver, but with one hoof slightly behind the other, neck stretched out with the chin lying between the legs, so that the nose comes an inch or two behind the hooves with the head lying along the legs and slightly to one side. So long as the lamb isn't too big – and that's unusual with naturally fed hill ewes – a ewe should have no trouble giving birth to a lamb that comes like this. Even if it's breach, but small enough, she will usually get it out backwards.

But the problems start if there's a leg back, or two legs back and the head has come out and swelled up so that you can't push it back down the canal far enough to hook the legs forward. Sometimes, you can have two little soft hooves visible (they are soft in the womb and quickly harden in the air) but the lamb won't come. These can be a front leg and a back leg of the same lamb, or a mix-up of legs from two or more lambs. You can only tell what's what by tracing the leg back from the hoof and feeling for the second joint. If it bends backwards it's a knee on a front leg and forwards it's a hock on a hind one. You may have to feel your way further down to where it's connected to the body to find out if both legs are connected to the same lamb. Once you have worked out what the problem is you have to manipulate the lamb into the correct birth position. This usually involves pushing it back far enough into the womb to create space to manoeuvre it. The ewe will be straining to push it out, so you have to wait and push back between contractions. It's fairly straightforward if there's only one lamb, and once it's straightened up the ewe will often be able to get it out on her own. But by the time you become aware that she's having trouble the lamb could be in difficulty and it's best to get it out as quickly as possible.

Sometimes the umbilical cord gets wrapped round the lamb while it's in the womb, starving it of oxygen. At other times if it is a big lamb it will be presented as a humped spine, because all four feet are pointing back into the womb. It's quite common when a Suffolk or other large meat breed has been crossed with a small ewe, which has become too fat during the winter, to have a lamb that is too big for the ewe to give birth to without help, because she quickly becomes exhausted. You have to wait until her time comes, catch her and pull the lamb out, working with her contractions. Often big single lambs result in 'hung lambs'. These are when the lamb is too big to come down the birth canal in the normal birth position. As the ewe's contractions intensify the head is forced out but the legs slip back to make way for it. Once the head is out it soon swells and unless you catch it early the lamb will choke to death and the ewe die of exhaustion.

If you're lambing on rough land where there are lots of places ewes can sneak off to have their lambs, sometimes the first thing you come across is a ewe with a grotesquely swollen lamb's head lolling out of her back end. If the lamb is still alive there will often be an obscenely swollen dry tongue protruding from its mouth. You can neither push it back nor pull it out because the head is too big and both front legs will be bent backwards. I once found a lamb hanging like this and in a desperate attempt to save its life, pulled on its head and got it out alive even though both legs were back. I didn't realise I'd dislocated its neck in the process and for the rest of its life it went around with its head at a quizzical angle.

Quite often the shepherd will be faced with a dead lamb. He usually finds the ewe standing over the carcase, defending her lamb from all-comers. It is best to leave her guarding it until he can find a foster lamb – either an orphan, whose mother has died, or one of a set of twins. If the twins are one of each sex he will probably leave the gimmer lamb with its mother so that the inherited instinct that attaches it to its home ground – the heafing instinct – can be passed on. If both twins are gimmers he will have to decide whether to take the bigger or the smaller. The bigger may be more likely to get over the shock of fostering, but if the ewe is doing a good job he will want to leave her with the better of her own lambs in case the fostering doesn't work. If they are both ram lambs he will leave the bigger, stronger one with his mother unless the smaller one is too small to get over the shock of being fostered.

The decision is a fine one – and has to be made quickly – but having chosen his lamb the shepherd proceeds to the next stage, which is to get the ewe to accept the strange lamb as her own. This is nearly all about how it smells. The knack is to disguise its smell so that the ewe believes it to be her lamb. Most ewes do not readily accept another ewe's lamb. Although they are fiercely protective of their own, they are usually hostile to the offspring of any other ewe.

The age-old, and still the best, method is to skin the dead lamb, to make a little jacket of the skin to fit the foster lamb, often with the mother standing close by to keep her interest. It helps sometimes to have the dog lying off, watching the ewe, just to keep her defences up. If it can be done right there, out on the hill, so much the better. This is where that other essential of the shepherd's trade comes in, a sharp pocket knife. The way to make the jacket is first to cut up the dead lamb's belly, from its anus to within about an inch of the throat, being careful not to penetrate the membrane of the abdominal cavity, and to leave enough at the neck-end to make a collar to fit over the foster lamb's head. Then cut round the legs above the first joint and pull the legs back through the skin. Then pull the whole skin off the lamb and over its head. If it is done well you will be left with a little jacket with five holes in it – one through which you can pull the foster lamb's head and the others for the legs. It's just like dressing a baby. Then, for good measure, the dead lamb's skinned carcase should be rubbed over the exposed head of the foster lamb to disguise its smell.

Then quickly pick up the carcase, leave the live lamb in its place and get out of the way as fast as possible. It helps if the live lamb is new-born and still finds it difficult to get to its feet because a foundering new-born always seems to provoke the ewe to greater maternal affection. If, as most ewes do, the mother accepts the imposter straight away then you will have no further trouble with the adoption. She will be over-joyed that her lamb has come back to life and nuzzle it making little bleating, mothering noises, encouraging it to suckle and licking its coat with small quick licks.

The new jacket will not come off if it is well-tailored and a neat fit. It needs to be left on until the ewe has completely accepted the lamb. If it is left on too long in warm weather it will start to putrefy, although that never seems to bother the ewe if she has accepted the lamb as her own. But if she hasn't been taken in and isn't too particular about the smell not being right – it's always the smell – you could try to disguise it with Johnson's Baby Powder or Eau de Cologne, or if all else fails Chanel No. 5 sometimes works.

I never found a ewe that cared a jot about the colour of the little jacket her lamb had been fitted out with. A black skin on a white lamb, or vice versa, worked fine so long as it smelled right. But, if she rejected the lamb at the outset, it could often be a long tedious job with them both cooped up in a pen until the ewe's milk had come through the lamb and it had taken on its mother's smell. Sometimes nothing would induce a ewe to take another's lamb and she would spend the summer geld, getting fat and often, it seemed, slightly unhinged by her loss.

Sometimes it worked if her head was secured in something

like the stocks, so she could stand up and lie down but not turn round to attack the lamb. But there is a type of fiercely independent ewe (often the best of mothers with her own lambs) that will actually starve herself to death, rather than accept a foster lamb. Such a ewe will either refuse to eat until her milk dries up, or stamp on the lamb whenever it tries to suckle, or in extreme cases, butt it against the side of the pen and kill it.

Other ewes are obsessive mothers, continually fussing over their children and hardly daring to let them out of their sight. This type would lick and paw the body of their dead lamb for a long time after birth, in a pathetic effort to get it to stand up. James Hogg tells of a ewe that stood over the body of her dead lamb for weeks on the open hill:

> It often drew the tears from my eyes to see her hanging with such fondness over a few bones, mixed with a small portion of wool. For the first fortnight she never quitted the spot, and for another week she visited it every morning and evening, uttering a few kindly and heart-piercing bleats each time; till at length every remnant of her offspring vanished, mixing with the soil, or wafted away by the winds.

There was also the occasional earth-mother type that never had enough lambs. She would entice away from their mothers any lambs whose maternal bond was slightly tenuous, such as the weaker of two twins, and allow them to suckle from her. I had one older ewe that had borne four sets of twins in previous years and was disappointed to give birth to a single lamb after her fifth pregnancy. She ranged round the lambing field and persuaded four more lambs, each one of a pair of twins, into believing she was their mother. It took ages to match them up again with their real mothers because it was hard to work out which twin belonged to which ewe when they were too young to have formed an attachment. The ewe never recovered from her perverted urge to steal other ewes' lambs, and she was a nuisance all summer.

The Hindhope flock only receives hay when the snow is too deep for them to forage. Surprisingly, a blizzard can be better than a heavy, even fall of snow, because a blizzard will blow the snow off the hill and leave areas exposed for foraging. Over the last few decades, since the ubiquity of the quad-bike, it has become easier to feed hill sheep in winter. But the traditionalists believe hill sheep should only be fed in the direst emergency, and some would say not even then. Those who routinely feed see no harm in it. In essence it comes down to an argument between the purists and the pragmatists, and they both have a point. For the purists hand-feeding undermines a flock's integrity, making it something less than a hardy hill breed, and destroys its independence. Breeding from sheep that have survived on hand-feeding tends to soften them, and create sheep dependent on expensive bought-in feed. If you're going to feed sheep with feed carried in from somewhere else you might as well not keep hill sheep at all. For the pragmatists it's simply about current profit. Taking a long view is all very well, but if you make more profit by feeding than not feeding there can be no contest, especially in a really bad winter when losses of hill sheep that are expected to fend for themselves can be very large indeed.

The Borders moors have long had some of the worst weather in Britain. Snow has been recorded, at the weather station on Eskdalemuir, falling on every single day of the year. James Hogg describes the terrifying storm that arose in January 1794:

> But of all the storms that ever Scotland witnessed, or I
> hope ever will again behold, there is none of them that
> can once be compared with that of the memorable

night between Friday the 24th and Saturday the 25th of January, 1794. This storm fell with peculiar violence on that division of the South of Scotland that lies between Crawford-muir and the Border. In these bounds seventeen shepherds perished, and upwards of thirty were carried home insensible, who afterwards recovered. The number of sheep that were lost far outwent any possibility of calculation. One farmer alone, Mr Thomas Beattie, lost seventy-two scores [1,540 animals] – and many others in the same quarter from thirty to forty scores each. Whole flocks were overwhelmed with snow, and no one ever knew where they were until the snow was dissolved, and they were all found dead. I myself witnessed one particular instance of this on the farm of Thickside: there were twelve scores of excellent ewes, all one age, that were missing all the time that the snow lay, which was only a week, and no traces of them could be found; when the snow went away they were discovered all lying dead with their heads one way as if a flock of sheep had dropped dead going from the washing. ... When the flood, after the storm, subsided, there were found on that place, and the shores adjacent, one thousand eight hundred and forty sheep, nine black cattle, three horses, two men, one woman, forty-five dogs, and one hundred and eighty hares, besides a number of meaner animals.

℃ 6 ℃

THE SCOTCH BLACKFACE

The day will come when the jaw-bone of the big sheep (caoirich mhora) will put the plough in the rafters; when sheep shall become so numerous that the bleating of the one shall be heard by the other from Conchra in Lochalsh to Bun-da-Loch in Kintail ... and henceforth will go back and deteriorate, until they disappear altogether, and be so thoroughly forgotten that a man finding the jaw-bone of a sheep in a cairn, will not recognise it, or be able to tell what animal it belonged to ... and the whole Highlands will become one huge deer forest ...

Coinneach Odhar, the Brahan Seer

I N THE NEWLY OPENED-UP HIGHLANDS MATTERS TOOK an entirely different course. The primitive crofters' sheep had been as unsuited to the new eighteenth-century imperative as were their keepers; no amelioration could have turned them into the kind of sheep that would pay the rents demanded by the Highland lairds of their commercial tenants. They were too small, too slow to mature, too ill-shaped and, ironically, too prolific, and had lost any hardiness they might once have possessed by being housed at night (to preserve their manure for the crofters' crops). They were simply not the ovine material needed to exploit the Scottish wilderness.

For that, a wholly different type of sheep was required and in the Scotch Blackface the new flockmasters found it.

At one time, the 'Blackie' dominated all the harder hills in Scotland. Its range was huge, from Northumberland, across into Ayrshire and Lanarkshire, Argyll, Perthshire, Ross and Cromarty, and far out into the Western Isles. It also adapted itself to the hard wet areas of Ireland and the moors of the West Country, where few other sheep could be as productive. In fact wherever the soil was thin and acid and the winters long and hard, millions of these sheep ate their way across hundreds of thousands of acres of mountain and hill grazing. Their output of wool and meat over the last two centuries has been nothing short of stupendous.

These are the 'short sheep with the lang thing'. And although Cheviot devotees will hardly be brought to admit it, the 'Blackie' is tougher than the Cheviot. Their wool is not as valuable as the Cheviot's, and they are slower-growing, but taken year on year the weight of sheep meat and wool per acre is probably not much different from that produced by the Cheviot. For sheer survival on the hardest of black heather hills, with little or no supplementary feeding, through some of the worst weather in Britain, these sheep have no rival. By about 1850 the century-long competition between the two breeds was settled by the Blackface establishing itself in the Central Highlands and Islands, where the Cheviot could not survive. The true difference between them eventually became clear; it was the quality of the grazing and not the severity of the climate that determined which of the two would do better upon a hill. And, unlike Cheviots, Blackfaces can live on heather.

Nobody is really sure where it came from, this horned, black-faced, active, shaggy-woolled, resilient sheep. It seems not to be part of the mainstream of ancient British sheep; rather it is a later arrival in Britain than the Celtic sheep. Some commentators have given it an exotic Argali ancestry, but apart from the heavy horns it is hard to see any resemblance to the Argali. The best guess is that it was of Saxon origin. There were certainly black-faced horned sheep in medieval Norfolk – corn growing on the hungry sandy Norfolk soils depended on their manure – and during the late seventeenth century and early eighteenth the same type metamorphosed into the Swaledale. It first spread from the eastern side of the Pennines into Westmorland and Cumberland, so that by the beginning of the eighteenth century it seems to have ousted most of the primitive sheep from the hills of Galloway and Ayrshire and contended with the Cheviot in south-west Scotland.

By the end of the eighteenth century it was so well settled in its terrain that the Westmorland farmer, in answer to where his type of sheep came from, could confidently reply, 'Lord Sir, they are sik as God set upon the land, we never change any.' From that time onwards the type spread so rapidly that by 1794 it was described as the *original* breed of sheep of Selkirkshire and became called the 'Linton' after the great sheep market at West Linton in Peeblesshire, and it was only a short time before a huge number of Blackfaces were settled into the Highlands. George Culley in 1807 described them as having large spiral horns, black faces and black legs, a fierce wild-looking eye and a short, firm carcass ... covered with long, open, coarse-shagged wool ... they are exceedingly active, run with amazing agility, and seem best adapted of all others to high, exposed, healthy mountainous districts.'

It is believed that the importation of Blackface sheep into the Highlands began in 1752. Their amazingly rapid advance followed hard on the heels of the Clearances, when it became profitable to keep sheep in those mountainous places which up to then had only been thinly stocked with crofters' black cattle and household flocks. There were many farmers from the Borders and plenty of skilled shepherds willing to travel north with their families for the work. By 1795 the breed was established in Ross in large commercial flocks and within little more than fifty years it had colonised more than half the Scottish hills; and its origins in the north of England seem to have been forgotten, because by the middle of the nineteenth century William Youatt wrote: 'Lanarkshire may be considered as the nursery of the black-faced sheep for the more northern counties [of Scotland].' How this sheep from the peaty limestone rocks of the Pennines, with no known crosses, adapted itself to thrive on the thin, wet, acid soils overlying granite, and the pure heather hills of the Grampians, is nothing short of marvellous.

As it has become widely dispersed subtle regional variations have emerged reflecting the differing soils, climate and, of course, the fancies of breeders. To the inexpert eye these variations are hard to detect. The Perth type is large-framed with a medium-to-heavy fleece, whereas the Lanark is a slightly shorter sheep with a closer, denser coat. The original Newton Stewart, or Galloway type, from the hills of the southwest of Scotland is a finer-boned, compact sheep with a short, thick, rain-resistant fleece; this is the type that is found in the Western Isles, where it is called the Lewis. It has the finest wool of all the sub-types, perhaps because it was crossed with the native Hebridean when first introduced, or it may simply be the effect of the soil, or probably a combination of both.

But it is not all plain sailing for Blackies. In this hard country, much of it strewn with granite boulders, like icebergs, whose red tips poke through the surface everywhere, Blackface sheep face many challenges. One is being poisoned by eating Bog Asphodel, *Narthecium ossifragum, ossifragum* meaning 'bone-breaking' and coming from the ancient belief that the bones of animals that ate it became weak and brittle. This might be because animals grazing acid soils are unable to get enough calcium. Ingestion causes a severe photosensitive disorder, variously called 'alveld' in Norway, 'plochteach', 'yellies' and 'head greet' in Scotland, and 'saut' in Cumbria. The plant is a member of the lily family and is commonly found in boggy places on moorlands on acid soils in Western and Northern Europe. Its yellow flowers are borne on spikes, about six inches high, during July and August. The flowers close up and turn a brilliant orange just before they fruit, giving the impression that the plant is still in flower. Eating it causes the animal's tissues to become sensitised to light, they lose their wool, their skin seeps fluid which dries into yellow crusts, and then the tissue dies and is sloughed off. They can die of liver failure. Sometimes lambs' ears curl up and drop off and in a bad season as many as 10 per cent of lambs can die. They can occasionally be saved by shutting them up in the dark for about three weeks, although it is not sunshine that causes the illness, but a build-up of light-sensitive toxins in the tissues which react with sunlight to cause tissue damage; by then it is often too late to reverse its effects.

Much of Scotland is also infested with ticks. There is an old hill shepherd's saying that 'sheep brush the hill', meaning that grazing sheep will attract ticks, which will then be killed by dipping. Where the sheep have gone the ticks feed on deer and other smaller wild mammals and birds, and proliferate without check. The explosion of the deer population has caused ticks to be carried into parts of Scotland where they were unknown before. From Scottish sheep farmers' point of view, 'the deer has destroyed Scotland'.

Towards the end of the eighteenth century, half a century after the Blackface began to colonise the Highlands, a number of enterprising farmers and landowners saw the opportunity of importing Cheviots from the Borders into Sutherland and Caithness. In 1792 Sir John Sinclair, of Ulbster, who later became the first President of the Board of Agriculture under George III, took 500 Cheviots up to his estate at Langwell in Caithness.

There had been intermittent acts of opposition to this new farming throughout the eighteenth century, but in 1792 open conflict broke out when some crofters were evicted to make way for the new sheep. Many landowners, who could see that the old communal society was no longer sustainable, hoped that sheep farming would form the basis of a sustainable new Highland economy that could co-exist with the old ways. But this was not a realistic solution because the old ways could not survive the profound onslaught of entrepreneurial capitalism that swept through Scottish society after 1750.

In Easter Ross-shire crofters were enjoined to mobilise and gather up the sheep by a melodramatic handbill posted on church doors throughout the region: 'That the curse of the children not yet born, and their generations, would follow such as would not cheerfully go and banish the sheep out of the country.' About 200 men responded and rounded up a large flock of about 10,000 of the hated sheep intending to drive them from the glen. But the authorities feared this was a harbinger of revolution spreading from France and mobilised the Black Watch to arrest the ringleaders and quell the upris-ing. The court dealt with them leniently by the standards of the times, which suggests there was some sympathy for their cause. 1792 is mythologised as *am Bliadhna nan Choirich Mora*, the Year of the Sheep, but it may well stand out because it was the only violent protest by disaffected crofters against the introduction of the hated English sheep.

The effect of the clearances is more obvious in Sutherland now that the sheep have gone and the land is virtually deserted. Spread out OS sheet 9 for Cape Wrath and, apart from the road in and out, there is nothing. Not even the ruins you would expect if farms and villages had been abandoned within the last 200 years. The people and their settlements have gone just as if they had never existed and now the sheep have all gone too. All that remains is the empty hills scarred with rocks and studded with tarns, and the wild beauty of the sea, its towering cliffs, white beaches and dunes thrown up by centuries of storms. Every feature on the map bears a Gaelic name, the melancholy ghosts of a dispossessed people in an empty land. Once they had been uprooted from their land, most crofters did not stay long in the new coastal places the landlords tried to settle them in. The New World beckoned with far more enticing prospects than eking out a desultory living in a place where the winters last for eight months and the summers sometimes never come.

Patrick Sellar, the Duke of Sutherland's agent, and an enthusiastic early promoter of the Cheviot in Sutherland, wrote that by 1820:

> 100,000 Cheviot fleeces were annually sent from Suther land to the manufacturer, and 20,000 ewes and wethers to the grazier; this annual extraction from the Alpine plants of 20,000 carcases of mutton and 100,000 fleeces of wool is indeed most wonderful.

By 1874 this annual extraction had reached a peak of 240,000 fleeces and carcases before its long decline in the face of the competition from cheap imports from the New World. The irony is that barely a century after the crofters had been replaced by the Big Sheep their descendants in the New World were sending back, to the old country, chilled lamb and fine wool that undercut the very men who had displaced their grandfathers.

Despite its apparent climatic disadvantages the imported Cheviots adapted well to their new home in the northern wilderness. Flockmasters found that if the ground was not stocked too thickly, their sheep could live off the land throughout the year. This was helped by the moors of Sutherland and Caithness enjoying one great advantage almost unknown outside hill sheep-farming circles, which is that in the second half of the winter, from about January or February, the draw-moss comes. Cotton grass, *Eprihorum vaginatum*, is known to hill shepherds as draw-moss because when they are eating it the sheep draw the tip of the emerging succulent leaves up the sheath-like stem from the rhizome beneath. It is a species of sedge that grows on wet acid soils across the hilly and mountainous parts of Britain. A cold winter will delay its emergence, but when it comes it can mean the difference between life and death for many sheep. Where it is available they come through the winter in better condition than those fed on hay or other foodstuffs, because the highly nutritious plant has an almost miraculous effect, acting as a tonic to in-lamb sheep that have subsisted on winter vegetation for three or four months with a growing lamb inside them. There is a saying amongst hill shepherds that a good draw-moss year is a good sheep year. And when the tips first emerge the whole flock can be seen 'working the moss' – spreading out across the boggy ground, busily drawing out the blanched stems from the peat and consuming them with relish.

Through domestic selection and the influence of the terrain and climate, the Cheviot in the north has evolved into a different sheep, the North Country Cheviot, and has even divided into two distinct strains: the Lairg type, from Suther-land and Wester Ross, and the Caithness type, a bigger, heavier sheep that reflects the better going it enjoys. These are long 'stretchy' sheep probably the longest-lived and most produc-tive hill sheep in Britain and much in demand from farmers on lower ground for crossing with Border Leicester rams to produce another hybrid, the legendary Scotch Half bred. These are prolific when moved to good land, and they consistently rear two lambs a year when mated with a meat sire such as the Suffolk or a Texel. They also have one of the best carcases of any hill sheep and tend not to lay down fat before their carcase has matured.

The North Country Cheviot's success partly derives from

one of the truths of plant and animal breeding, which is that foundation stock always does better if it is moved from high to lower ground and from the pole towards the equator – north to south in the northern hemisphere. If an animal or plant (British seed potatoes are a good example) is reared in the harsher climate of the north of Britain, particularly in the hills, then it will perform better if it is moved downhill and south. With our large range of climate and topography we are well-placed to take advantage of this phenomenon. That is why Cheviot draft ewes from the far northern hills of Scotland have been in such great demand for over 200 years.

There are few places further west on mainland Britain than Cape Wrath, which is on nearly the same longitude as Penzance. But that is where the climatic similarity ends. In Cornwall the daffodils flower in mid-January; in Sutherland they do not emerge for another four months. Even in mid-May you still need an overcoat when people in the south of England are brushing off their garden furniture. For although the harshness of the northern climate is tempered by the Gulf Stream, the winters are long and for two months either side of the solstice the sun barely rises above the horizon. There is not much snow – at least on the lower ground – but there is not much shelter from the gales that sweep in from the Atlantic across the mountains and the wide Flow Country.

Balnakeil, 'the village of the church', is the last estate in the north-west of Scotland before Cape Wrath. Once the summer residence of the bishops of Caithness, it is one of the most valuable and productive sheep farms in Sutherland, one of a diminishing number still fully stocked and so remote that it can take five hours to make the hundred-and-twenty-mile round trip to Lairg along a single-track road with passing places. The underlying rock on the northern part of the farm is limestone and, as the old saying goes, limestone makes bone. It also makes the huge white dunes and sands of Balnakeil Bay.

Andrew Elliot's great-grandfather was one of the enterprising Borders farmers who leased the 20,000-acre estate from
the Duke of Sutherland in the early nineteenth century, shortly
after the railway reached Lairg from Inverness. It was a considerable journey to reach his northern enterprise. He had to
travel from his farm in the Borders by pony and trap the thirty
or so miles to Edinburgh, then by train via Inverness to Lairg
and then again by pony and trap sixty miles by Loch Shin and
Loch More, through the wild Sutherland hills, to Balnakeil.

In its heyday Balnakeil employed eight full-time shepherds
to manage 4,000 ewes. Most of the year's income came from
the great annual cycle of sheep sales at Lairg, once the largest
sheep market in Europe, where sheep from Balnakeil consistently topped the market. The sales began in August with
100,000 wether lambs straight off their mothers on the hill
and sold through two sale-rings throughout one long August
day. The lambs were sold in scores (twenties) from pens that
covered the hillside behind the mart. Buyers knew the improving potential of these sheep and came from all over Britain.
The most mature lambs could be transformed into prime
butchers' animals simply by moving them to better pasture for
a few weeks; while the smaller ones would be fed on root crops
in the lowlands and sold during the winter.

After the lambs, in September, came the sales of draft
ewes, when tens of thousands of four- and five-year-old ewes,
with all their lives ahead of them, were sold to farmers on
lower land to breed the Scotch Halfbred from a cross with the
Border Leicester. Later in September the ram sales attracted
the throng of farmers and shepherds from all the lonely
places in Sutherland and Easter Ross, taking a last chance to
get together, do some business and renew old acquaintance,
before the shadow of the long northern winter once again fell
across this bleak country.

Until comparatively recently all the sheep that left the hills of Sutherland travelled on their own four feet. It took the shepherds and their dogs over a week to drive the Balnakeil sheep to Lairg, stopping off every night in places where the sheep could graze and be safely contained. Hamish Campbell was head shepherd at Balnakeil for twenty years until he retired in 2010 on the 'traditional day', i.e. 24 November: 'The sheep's feet had to be right to walk that far without losing condition – as had the shepherd's. We took great care of our feet, with the best sprung boots and wool socks. Our tweed breeks buckled just below the knee with plenty of cloth to hang well down the calf, so that when it rained the water simply ran off.' A shepherd was paid twice a year. And by the time he had settled his tailor's and his boot maker's bills, as well as the other tradesmen he owed money to, there was little left. 'He just started all over again until the next six months had gone by.'

Now the huge sheep flocks have mostly gone from the Highlands and are rapidly going from Sutherland and Caithness. It seems that the Brahan Seer's prophecy is being fulfilled as the land becomes wilderness, the life slowly draining from it. There are even suggestions of introducing wolves and beavers, last seen many centuries ago. On present trends it will not be much more than a decade before the flow of sheep meat and wool dries up and the huge acreage is lost to agricultural production. This is probably the first time in British history that agriculturally productive land has been deliberately given back to the wilderness. But if sheep keeping ceases in the Highlands it is hard to see the clans returning to take undisturbed possession of the lands of their ancestors, as the Seer foretold. The most likely outcome will be that nobody will live here and nothing much will be produced. It seems perverse, to say the least, that humanity should retreat from half of Scotland and abandon such a resource of meat and wool.

There was an almost palpable *fin de siècle* feeling everywhere I went in Sutherland. Farmers and shepherds are growing old and retiring, with no young people to take their place. The solitary shepherding life no longer appeals to a sedentary and gregarious youth seduced by the bright lights of the towns, and unbroken to manual labour. All over Britain sheep are leaving the hills, but nowhere is the loss felt more acutely than in the northern counties of Scotland, where the income and social structure depend to such a large extent on sheep farming. Steadily and inexorably a whole way of life is drawing to a close: the camaraderie, competitive rivalry, collective gatherings and shearings – they all become a memory that lives only until the last one to remember dies.

All manner of things are to blame: the price of sheep for the last two or three decades; the EU paying farmers not to keep sheep; the difficulty for young people to get a start in farming because few tenancies are available; the shortage of labour. Whatever it is, it seems irreversible and it is tempting to see it as part of a wider decline that is causing us to abandon our remoter places, to retreat from the edges to a comfortable centre. Whatever the causes, there seems little doubt that two and a half centuries of commercial sheep-farming in the Highlands is drawing to a close. If all agricultural financial support is removed, or, more likely, the payments become social subsidies, the end cannot be far away.

7

THE LEICESTERS

'You can't breed rats out of mice.'

Jim Brown, Border Leicester breeder from Mindrum

ROBERT BAKEWELL'S AIM TO CREATE A BREED THAT would provide meat for the toiling industrial masses did not quite work out as he intended. As we have seen, his New Leicester was not a success as a pure breed and in the form that he bred it has virtually died out. Its direct descendant, the Leicester Longwool, is so far out of favour that it is on the 'endangered' list of the Rare Breeds Society, with fewer than 500 breeding females left in the hands of a few faithful breeders.

But the value of the Dishley Leicester lay in its legacy, because its descendants are the Longwool breeds that form the male half of the first cross in the sheep pyramid. There was something about the animals that Bakewell chose to breed from, some characteristic, that has endured in their cross-bred offspring; it gave almost every breed they were crossed with a superior carcase, earlier maturity, more lambs, and a greater vigour and fitness to survive. In short, they tended to imbue their progeny with just those qualities that made them suitable for meat production. And such was the fashion that there was hardly a breed or type of sheep that did not receive its share of the dominating New Leicester genes, with the result that nearly every sheep in the Western world has some Leicester blood, however dilute, running through its veins.

Many breeds did not derive any lasting benefit from this infusion; for some it was little short of disastrous, ruining their wool, making them run to fat before they had grown to full size, or turning frugal sheep into high-maintenance prima donnas. Such was the faith in its enhancing powers that many breeders believed that simply introducing Leicester blood to their own indifferent animals would have an almost magical effect, transforming them into a superior type by some form of alchemy.

However, in a way that Bakewell could hardly have foreseen, his New Leicester's legacy has proved to be of the most enduring importance to modern fat-lamb production because the modern British sheep pyramid owes its existence to his work. For not only did a cross with the Dishley Leicester transform the old Longwool types (not then breeds) into the modern Longwools – the paternal side of the first cross – but it also gave the Down rams used for the second cross many of their enduring qualities.

The four most prominent modern descendants of the New Leicester are the Border Leicester, the Bluefaced Leicester, the Wensleydale and the Teeswater, all of which are hugely important to modern British sheep production. Each breed has been developed to have qualities that nick with its corresponding hill breed, and which bring out the best of the hill ewe in its hybrid female offspring.

The Border Leicester goes back the furthest and owes its existence to two of the great eighteenth-century pioneers of English farming and Bakewell's most well-known pupils, Matthew (1731–1804) and George (1735–1813) Culley. They were the sons of a farsighted yeoman farmer from Darlington in Co. Durham, who was well enough off to send them in 1763 to be pupils to Bakewell to learn the new farming and stockbreeding and maintained a close friendship, bordering on hero-worship, which lasted to the end of Bakewell's life. They had been on the yeoman farmer's version of the Grand Tour – a trip around Europe to observe and absorb its agriculture – and upon their return, in 1767, took a lease of Fenton, a 1,100-acre farm in Glendale, in the fertile valley of the River Till in north Northumberland. The Culleys rode the English eighteenth-century agricultural revolution and made a fortune from progressive farming, ending up as substantial landowners. When they moved to Northumberland they not only imported the new optimism that gripped English farming but, more importantly for the future of English sheep breeding, they brought their stock of Dishley Leicester sheep.

This was a time of astonishing opportunity for commercial farming in England. It is hard to exaggerate the optimism, raw spirit of enterprise, and sheer effort that was poured into it. The stage had been set by a century and a half of protection from foreign imports, beginning in the second half of the 1600s with the first Corn Laws. Guaranteed prices boosted home production, and encouraged the landowning and farming interest to invest and innovate. Relatively high food and produce prices, throughout the eighteenth century and first half of the nineteenth, produced the money that paid for the enormous investment needed to transform British agriculture.

During these good times, millions of acres of what had been unfenced waste, or common, were enclosed, underdrained and brought into cultivation and the great landed proprietors consolidated their power by buying out thousands of smaller proprietors, and then let their enlarged estates to business-minded tenant farmers who could pay the highest rents and together become their partners in the great agrarian enterprise upon which their wealth and power were based.

During the first sixty years of the eighteenth century, the English countryside was ruthlessly transformed from the place where most people lived into an extensive, almost completely industrial system to supply the burgeoning urban demand for food. Farming became another mercantile endeavour and a source of profit to the nation. By 1764 England and Wales produced more cereals and potatoes per head of the population than at any time before or since.

It was not only the small proprietors who suffered; the village labourer was also dispossessed of his small inheritance in the open fields and on the commons that cushioned him from starvation when work was scarce and gave him a measure of independence. He was uprooted from the land and paid starvation wages, losing any direct interest in the produce of the soil. This huge rural labouring class suffered much misery and insecurity, and were severed from their roots on the land; this opened up a gulf in understanding between town and country, which endures to this day.

The Border Leicester is a product of this revolution. The Culleys most likely crossed the Bakewell sheep they took with them to Northumberland in 1767 with the local sheep they found there. It is not clear whether these were Cheviots or a longwoolled local type, called the Mugg, or another local woolly sheep, the Bamburgh (now extinct). A *mug* is a thick lock of wool that grows on a sheep's head and often flops over and obscures its face. Before the emergence of distinguishable breeds, many of the big woolly types were collectively known as Muggs.

There is, however, a less plausible suggestion that its creation was the work of John Edmistoun of Mindrum and a few other progressive Borders farmers, who in the mid-1750s travelled into Lincolnshire to buy 'improving tups' for their flocks

1. Hardy little Black Hebridean wethers, belonging to the northern short-tailed group of primitive sheep that once would have occupied the western seaboard of the British Isles, even into the Channel Isles. All primitive breeds are highly efficient at extracting energy from almost any kind of vegetation, but the Hebridean is in a league of its own.

2. *(Left)* Extravagantly horned ram of the Manx Loghtan (*lugh dhoan*), the national sheep of the Isle of Man. These are proud sheep, unconquered by modernity. The horns are more ornamental than useful as rams can split their skulls if they start fighting – which they find irresistible.

3. *(Right)* Castlemilk Moorits were created in the last century by Sir Jock Buchanan Jardine to grace his park at Castlemilk in Dumfriesshire, with the practical benefit of providing wool to clothe his estate workers. They are a cross between a wild Mouflon, Manx and Shetland, and are a unique modern manifestation of an ancient breed.

4. Jacobs come with varying numbers of horns, up to six, and all with unique piebald fleeces, 'the sportings of nature, speckled and spotted,' unchanged since The Creation, a direct link with the ancient world.

5. Portlands are another breed with an ancient lineage stretching back into the Iron Age. Their last redoubt was on the Isle of Portland in Dorset, but they would once have been ubiquitous across south-west England. This proud little ram with his impressive horns, is considering whether or not to charge at me.

6. Portland lambs are born foxy ginger, grow a creamy fleece as they mature but never lose the tell-tale tan colouring on their legs and dished faces. Note the characteristic black line running through the outer edge of the ewe's horn.

7. Another remnant of the northern short-tailed group, North Ronaldsays live entirely on seaweed for two-thirds of the year, where they are confined to the foreshore of their island by a wall built above high water, to 'louping height.' Not all the sheep treated the wall with the same respect – these two used it as a look-out post.

8. Picking the best. A group of North Ronaldsays graze the fresh, crisp blades of ware recently exposed by the retreating tide. They lie up, on the beach head, chewing the cud at high water, then, when the tide turns they follow the ebbing water, competing for the fresh seaweed. The more intrepid will sometimes swim to a rocky outcrop for the freshest fare.

9. Some people will go to any lengths to grab a headline! Louise Fairburn in her wedding dress, made from the fleece of one of her Lincoln Longwools, holds Risby Olivia, her champion ewe. The men wore woolly waistcoats and the guests dined on her Lincoln lamb.

10. Risby Ruby and her lamb. The names of pedigree sheep begin with a different initial letter of the alphabet for each year. Every part of a Lincoln Longwool, except their black Roman noses and ears, grows ringlets of the highest quality lustre wool.

11. Traditionally managed hill flocks are still lug marked to indicate ownership. This Cheviot ewe from Tim Elliot's Hindhope flock in the Borders is marked with an upper key bit and lower fold bit in its right ear. Note the black nose, one of those secondary characteristics, like black cleats (the cleft between the hooves) that is a sure indicator of hardiness.

12. Swaledale ewe with her new-born Mule twin lambs. This first cross in the sheep pyramid, with a Bluefaced Leicester ram, produces an outstanding hybrid breeding female which goes on to be crossed a second time with a Down ram to provide much of the lamb we produce for the table. Note the intensely contrasting and stylish black and white colouring on the lambs which they have inherited from both their mother and father.

13. Robert Bakewell (1725–95), on his cob, at the height of his powers. During many summers he tirelessly criss-crossed England on horseback noting the effect of his New Leicester rams on the hundreds of flocks whose owners had hired them. He is the most prominent of those breeders who began the transformation of English sheep from wool to meat producers.

14. Native to Sussex and first improved by John Ellman, the Southdown was taken up by many nineteenth-century breeders. One of the most successful was Jonas Webb of Babraham in Cambridgeshire. The drawing is propaganda, caricaturing Webb's sheep as preposterously meaty carcases with impossibly fine boned legs.

15. By 1840 the Scotch Blackface had almost entirely supplanted the old Scottish breeds and come to dominate all the harder hills of Scotland. This shearling ram is of the Perthshire type, which has a large frame and heavy coat. The wire stretcher training the horns into an elegant shape (like a dental brace) can just be made out behind its right horn.

16. Caused by the differences of climate and terrain, over time, subtle variations in the Blackface breed arose. This ewe with her ram lamb is of the Lanark type, dominant in central and southern Scotland and the Borders. Its fleece is shorter and denser than the Perthshire type

17. The Wensleydale has lost out in recent years to its ancestor the Teeswater and its more socially ambitious cousins the Bluefaced Leicester and Border Leicester, but it still has a devoted following of breeders who want a fine crossing sire for horned blackfaced hill ewes, particularly the Dalesbred. It passes on to its hybrid offspring, the Masham, its fecundity and heavy lustre-wool fleece, pirled to the end of each ringlet.

18. Wensleydale ewes are excellent milkers and well-able to rear the twins and triplets they usually produce. The intense blue skin, inherited from their progenitor 'Bluecap' is evident, even from birth, in the faces and ears of these new-born lambs.

19. 'It should have a head like a solicitor.' Bluefaced Leicesters were first bred around Hexham for crossing with Swaledales and Blackfaces to produce hybrid Mule breeding ewes. Once considered the poor man's Border Leicester, so popular and successful have they become that only buyers with deep pockets can now afford the best of them.

20. Buyers at the biggest ram sale in Europe held each September at Kelso, lining the ring to admire 'the cock o' the lug and the glint o' the eye' of this fine Border Leicester ram lamb. This breed is Bakewell's enduring Borders legacy.

21. The breeder of this pen of rams has heeded well the Suffolk Society's breed standard that their 'hind legs should be well filled with meat and muscle'. The Suffolk vies with the Texel to be the most ubiquitous meat breed in Britain, the sire of many millions of prime lambs that grace our tables every year.

22. You wanted meat, well here it is! Beltex rams at Kelso. Uncompromising examples of the breeder's art, like a gang of night-club bouncers.

23. Herdwicks are the tough mountain sheep native to the English Lake District. They are said by their devotees to 'live on fresh air, clean water and good views', as can be seen from the sparse pickings that sustain these sheep in January at Buttermere, in the heart of their ancestral domain.

24. Herdwick ewes straight off their heafs in October, at the annual draft ewe sale at Cockermouth. The ewes in the foreground have been rudded to dress them up for sale, whereas those in the pen on the left of the picture prefer the natural look.

25. The golden hoof. A Dorset Horn chilver that has spent the winter folded on a crop of turnips and kale. Note the white flesh of the stumps of turnips that the sheep have eaten flush with the surface and the black deposits of sheep muck spread evenly over the land between the flints that litter this thin soil

26. Tess, the author's Border Collie cross Bearded Collie bitch. She is typical of the working sheep dogs of the north of England.

27. There are more ways to move a flock of sheep than barking at their heels. A kelpie, 'mounted up' and running across the backs of a flock of Merinos.

28. *(Left)* East Friesland milking ewe at Orchid Meadow Farm in Dorset. There is a genetic correlation between heavy milking and a slender frame. These sheep are the ovine equivalent of dairy cows with their delicate feminine pink skin and fine bones.

29. *(Right)* A batch of East Friesland sheep being milked in the parlour at Orchid Meadow Farm. Each batch takes about ten minutes to milk; their feed is rationed according to milk yield and dispensed through the white tubes above their heads.

30. Texel ewes. Short-necked, square blocks of meat. Bakewell would have been impressed that modern breeders have succeeded in making the hindquarters bigger than the forequarters, something that eluded him.

31. This Whitefaced Woodland ram, from Woodland Vale in the Derbyshire Peak District, is a relative of the Welsh Mountain, the Herdwick and the Cheviot and part of that great affinity of hardy sheep that was once native to the west side of England in a great sweep from the Scottish Borders to the West Country.

32. The quintessence of chalky whiteness, the Wiltshire Horn has long been acclimatised to the calcareous downs of its eponymous county. It is our only breed that naturally sheds its pelt (hardly a fleece) in summer. This ewe survived entirely on grass all winter and had just given birth to twins a quarter of an hour earlier.

(probably Cheviots). They brought back a big heavily woolled type of old Lincoln from the Lincolnshire Wolds, almost certainly of the same medieval Lincoln breed that Bakewell used to create the Dishley Leicester. These Lincolns would have been crossed with the native Cheviots and it is probable that the sheep that eventually became the Border Leicester was created in this way.

The Culleys' clear purpose was the same as Bakewell's (and every other livestock improver's at the time), to breed a more profitable type of meat sheep to supply the growing urban market. And it is more likely that the Border Leicester is the result of a cross between a Cheviot and a New Leicester, with or without some other Longwool blood. At some point it lost the long, lustrous wool on its father's side and developed a shorter, tighter fine fleece, more like the Cheviot than the Leicester, and it also inherited the Cheviot's distinctive cock o' the lug. It is a tall, powerful barrel-shaped sheep designed to breed female hybrids by crossing with Cheviot and other hill ewes.

It is almost certain that the Culleys adopted Bakewell's incestuous breeding methods to create their new breed: 'in-and-in' breeding (fathers with daughters, mothers with sons, and siblings with siblings) until the characteristics of the new breed were fixed. Defective animals would be culled – and eaten. They kept the flock closed (did not buy any rams or sell any females) to preserve the purity of their bloodline and almost always refused, *pace* Bakewell, to sell their rams – preferring to hire them out, often at huge prices.

The following extract from George Culley's letter to his brother Matthew on 24 December 1784 succinctly explains the difference in the meat between the old type of sheep and the new sheep they were trying to create. Having ridden through Norfolk, Culley wrote:

Much is to be done in these horned countries, but when I
know not, the change is so very great that people cannot
easily reconcile their ideas to a long wooled [sic] sheep
with a thick carcase, who have been used to their deer
kind of sheep, the more they approach to deer, the more
lean flesh and less fat in proportion but more gravy of
the claret kind, Mr Matt. Mr Bakewell and I have recommended this consequently gentlemens [sic] mutton and
the manufacturers [sic] mutton are of two kinds, the one
lean with gravy, the other fat with oil, the gravy meat
open grained and porous, the fat meat fine grained and
close.

The 'deer kind of sheep' is a reference to the predominant
Norfolk breed which was one of the old types that Bakewell and
his disciples thought 'profitless', and 'vessels to carry manure
from one field to another', scathingly dismissing their value
to the old Norfolk fold–course system, by manuring the land
for the next crop of corn. But perhaps more telling is the way
Culley's words aptly illuminate the new mercantile approach
to farming. The profitable sheep he breeds provide 'manufacturers mutton … fat with oil', firmly to be distinguished from
the lean flesh of 'gentlemens' sheep that gives 'gravy of the
claret kind'.

To be fair to the traditionalists, who formed the majority of their neighbours and who took a principled dislike to
the Culleys' fat 'manufacturers' sheep, with some reason they
feared the soft Leicester would adulterate the hardiness of
their native breeds. One neighbouring landowner, William
Mure, hired a ram from the Culleys, only to find that his
neighbours, who were at that time inveterate enemies to the
Culley sheep, had sabotaged his effort, by introducing 'a black-
guard Galloway ram' amongst his ewes and out of the whole
flock there were only five ewes that 'had not got the ram …'

This resistance to the Culley Sheep was not, as it might have appeared, just local prejudice against in-comers. The detractors had a good point that they were nothing much as a breed in their own right: they ran too early to fat, their flesh was tasteless compared with the old breeds, they struggled to average more than a lamb and a third from each ewe, and they were too soft to subsist on ordinary grazing, needing to be fed throughout the winter (on the relatively new crops of turnips or cabbage, or on hay and corn).

Gradually commercial reality brought round the Culleys' critics to their finer-boned, barrel-shaped Leicester. As their sheep became more sought-after, the brothers formed the Millfield Society, along the same lines as Bakewell's Dishley Society, or 'Ram Club'. It was not long before farmers and landowners from all over the north were hiring tups from the Culleys (sometimes even from Bakewell himself) and the Culley Sheep established a formidable reputation throughout the livestock-farming districts on both sides of the border.

After Bakewell's death in 1795 the northern breeders of the Dishley Leicester began to distance themselves from their southern confrères, and gradually the Culleys' version of the Leicester developed into the distinctly different breed that has come to be called the Border Leicester. It became apparent in the early nineteenth century that this Border Leicester when crossed with native hill ewes, particularly the Cheviot, made a fine hybrid, which by 1850 had become renowned as the Scotch Halfbred and became the principal sheep export from the Scottish Borders. It has unrivalled maternal instincts, and when crossed with a Down ram is capable of rearing a set of twins to the same size as their mother within four months. This is hybrid vigour in action and is Bakewell's and the Culleys' enduring legacy.

The great success of the breed is proof of the feeling for livestock that is the defining characteristic of Borders stock farmers. It is no coincidence that the largest ram sale in Europe takes place at Kelso every year in early September, when thousands of rams from all the Longwool and Down crossing breeds are dressed up for sale in their smartest clothes and offered for sale to buyers from all over the country.

In the 1950s the Border Leicester expanded its appeal into the Welsh Marches, where, crossed with Welsh Mountain ewes, it created another great British hybrid, the Welsh Half-bred. This crossing proved nearly as successful as its Scottish cousin, and at the height of its popularity hundreds of thousands were bred each year for selling on to lowland farmers to cross with terminal sires. This trade has declined considerably in recent years, but the hybrid still forms a successful and valuable part of the national flock.

Modern pedigree breeders of highly valued sheep, such as the Border Leicester, use all the latest scientific breeding technology: embryo transplants, artificial insemination, fertility treatment, ultrasound scanning and frozen semen. As a result it is now possible to have two ram lambs from the same ewe by different mothers and the same or different sires: one could be a fresh embryo and the other a frozen embryo from a previous impregnation, both implanted in a donor ewe, using the same techniques as in modern human fertility treatment. Stores of deep-frozen semen from prize rams can be held at very low temperatures in liquid nitrogen and kept viable for decades. It is no longer necessary to export live rams for breeding. And now that they can be milked of their semen, the best rams make proportionately higher prices. One successful Border Leicester breeder I spoke to also practises 'line breeding' – the same incestuous method that the Culleys copied from Bakewell. He wryly summed it up: 'when it works it's line breeding and when it doesn't it's in-breeding.' In livestock breeding, of course, you (or someone else) can always eat your mistakes.

It is becoming common practice to flush out huge numbers of eggs from one particularly good ewe, fertilise them with different rams and implant the fertilised eggs into donor sheep. The same pedigree breeder 'harvested' eighty-nine embryos from flushing five ewes in 2010. He didn't use them all, but it shows how much quicker an animal's breeding potential can be evaluated now than in Bakewell's day. Pedigree breeders don't have to hire out their rams and sneak meanly around the countryside spying on their progeny over hedges.

Ultrasound pregnancy scanning of ewes is now routinely used in sheep farming, and is exactly the same as practised in ante-natal departments across the world. Opinion is divided between those who scan to find out how many lambs each ewe is carrying, so they can feed accordingly, and those who don't see the point because they feed according to each ewe's condition no matter how many lambs she might be carrying. There is also a romantic view that scanning is a waste of money and an unnecessary upset for the ewes, and knowing exactly how many lambs the flock is going to have takes the fun out of lambing. It is also too tempting to count your chickens before they've hatched. As some lambs are bound to die, the shepherd can only ever feel he has suffered a loss.

'It should have a head like a solicitor,' explained one old Bluefaced Leicester breeder when I asked him what a good one looks like.

Until after the last war, the Bluefaced Leicester was almost unknown outside its home territory in the country around Hexham in Northumberland. But over the last fifty years or so it has spread the length of the British Isles and rather eclipsed its Border Leicester cousin. These are such odd-looking sheep that it is hard to see why anybody would want to keep them, especially when you take into account their propensity for dying on the slightest pretext. In fact, keeping them only makes sense in the light of their hybrid progeny. They are the most specialised and refined Longwool crossing breed, developed at the beginning of the last century especially for crossing with black-faced horned sheep, particularly Swaledales, (as we have seen), to produce the hybrid Mule.

At first they were considered to be the poor man's Leicester because they were thought inferior to the Border Leicester, and their progeny inferior to the Scotch Halfbred. But things have moved on since then. If anything, Mule gimmer lambs are more sought after than Scotch Halfbreds and the best Bluefaced Leicester rams are now beyond the reach of paupers.

It is astonishing that the early breeders could see that such an odd sheep would match so perfectly with a Swaledale. Tall, long in the body, with blue-black colouring to the skin, and forwardly erect blue ears, Bluefaced Leicesters are the only breed that often sit with their heads back and their Roman 'solicitor's' nose in the air. The ewes are very prolific; on average every ten ewes will rear twenty-five lambs, five having twins and five triplets. Such prolificacy is remarkable even in Longwool sheep. Their fleece is sparse, 1–2 kg of demi-lustre wool, and much finer and shorter than the Border Leicester's; in fact there is hardly enough wool to keep them warm. But in combination with the Swaledale, the lustre wool of its father and the heavier fleece of its mother result in a high-quality, much heavier lustre wool-fleece. This is only one of the more obvious ameliorating effects of the first cross. Just why it should produce such an outstanding breeding sheep is one of the mysteries of ovine genetics, because when crossed the other way round – Swaledale ram onto Bluefaced Leicester ewe – the result is disappointing.

Those who created the Bluefaced Leicester did so by a combination of instinct and practical experiment. There were no geneticists involved; they undertook no scientific research;

and yet by knowing what they were aiming for, by trial and error and an almost Bakewellian understanding of the qualities that lay within, they created a sheep that has transformed the value and productivity of draft Swaledale ewes and the quality of British lamb for the table.

Bluefaced Leicesters are sheep of the 'Shire' – that part of Northumberland south of the River Tyne that would, but for historical accident, belong to Co. Durham. The land has been settled far longer and on a more human scale than the straight lines and huge rectangular fields in the rest of the county across the Tyne. Bright streams splash down wooded denes between small hilly fields and narrow roads tunnel in diffuse green light through overgrown hedges, more like the lanes of Somerset than the wild country to the north or the windswept moors in the south. This is classic livestock-rearing country and the home of the Bluefaced Leicester. The flocks are not large: thirty pedigree ewes is a respectable size. Most commercial farmers haven't the patience to bother with pedigree sheep. So the whole thing depends on a few dozen specialised breeders who maintain the integrity of the breed.

How do they decide what they ought to look like? Do they breed them how they fancy they ought to look, or are they aiming for some objective ideal standard? Most breeders are no better at describing what they are aiming for than the one who thought they should have a head like a solicitor. I know that much of livestock breeding is instinctive, but surely there must be some Platonic version of the Bluefaced Leicester against which they judge their stock? If you ask them breeders will point out one of their sheep as a 'good one', but it is almost impossible to see why it should be better than any of the other sheep standing next to it.

One of the best breeders, and a respected judge, said he usually only sees the faults in a sheep; 'nothing catches my eye like a defect,' he told me, 'but I take their virtues for granted'.

He explained that he carries in his imagination a picture of the ideal Bluefaced Leicester against which he judges the real ones. And the best breeders spend their lives trying to create this ideal. But where the ideal comes from is almost impossible to say. And just because a ram is a good specimen does not mean it will make the highest price at auction. Breeders will sometimes buy a ram with qualities that might correct some defect in their ewes, always keeping in mind the ideal sheep they are trying to create.

In such a small world as pedigree breeding, where nearly everything is sold at auction, if the leading breeders appear interested in a particular sheep that will tend to push up the prices. So they try to avoid being seen bidding for fear someone will 'run' them – bid up the price without intending to buy – either because it is believed that the ram must be something special, or just to have some fun and see how far the bidding can be pushed before someone chickens out.

They usually get someone else to bid with instructions not to go beyond a certain price, while the real buyer sits back and watches. Breeders will often spend the money they have made selling their own rams by buying someone else's rams. It is not unknown for the auction price to be manipulated to enhance the seller's reputation by trying to 'top the market' and make a name for himself and his flock. One well-known trick is for the seller to make an agreement with a buyer that he will refund a proportion of the hammer price either because he has agreed to keep a share of the animal or simply in return for the buyer pushing up the price. This is a fraud, and would invalidate the sale, but proving it is another matter.

The Bluefaced Leicester has now divided itself into two types: the 'fancy' kind, which vies with its once grander cousin, the Border Leicester, for crossing with Cheviots and other white-faced sheep like the Welsh Mountain; and the commercial type supposed to be better for crossing with Swaledales to breed a Mule hybrid with strongly contrasting dark chocolate and white markings. It is hard to fathom why such starkly contrasting markings are so popular with breeders crossing with the Swaledale. There is nothing to say that sheep with such colouring make better mothers or have more and better lambs. It's just a fashion. But it's a fashion driven by the market for lamb meat, which the pure Bluefaced Leicester and its Mule offspring, and her lambs, are created to satisfy.

Another variation on the Leicester theme is the Wensleydale, whose crossing mate is a cousin of the Swaledale, the Dalesbred, naturalised on the moors of Upper Wharfedale and Nidderdale. Dalesbreds are a good example of domestic selection varying a breed over time. They are similar to Swaledales, with the same short, loose fleece, long legs and active nature, but have been selected for their sharper black and white colouring. They cross well with the Wensleydale to create a breeding female, the Masham, named after the North Yorkshire town and a rival for the Mule. The Masham has a heavier, more lustrous and pirled (falls in ringlets) fleece and is reckoned to be more prolific than, but not as hardy as, the Mule.

The Wensleydale is one of the few pure breeds whose ancestry is known for certain. A couple of hundred years ago, in the lower parts of the Yorkshire Dales, particularly Wensleydale and Teesdale, the indigenous longwoolled sheep were called Teeswaters, members of the great family of English Longwools, probably with a common Roman ancestor. David Low described them in 1839 as

> the most remarkable of the inland breeds ... so named
> from the valley of the beautiful river which separates the
> counties of York and Durham ... The breed extended with
> some change of characters, northward into Durham and
> southward through the greater part of Yorkshire, until it

merged in the heavy-woolled sheep of the marshes (Lincolns) on the one hand, and those of Leicestershire on the other.

These Teeswaters were probably the biggest sheep in Britain in the eighteenth century: tall, lean, large in the frame, with a heavy lustrous fleece, up to 24 lb, and second only in quality to that of the ancient Lincoln. But they had 'an exceedingly uncouth form, with coarse heads, large round haunches and long, stout limbs'. They required the best of pasture, like cows, and could not survive the winter without supplementary feeding with hay and corn. Well-kept they were most prolific, usually bearing twins and triplets, and were fine milkers, capable of suckling and rearing all the lambs they produced. They took three or four years to grow to full size, but that did not matter because their huge carcase was never intended to be eaten, or at least not until it had produced a respectable amount of wool. One animal, killed for Christmas 1779 at Stockton-on-Tees, had a dressed carcase weighing 17 stones 11 lb, plus 17 lb of tallow. Alive it would have weighed about 40 stones, about the same as a Shetland pony, and probably as edible.

As the Bakewell influence spread, Teeswater breeders recognised the limitations of their local sheep as meat sheep and began to introduce Dishley Leicester rams into their flocks to try to breed a better mutton carcase. One Richard Outhwaite of Appleton paid 40 guineas in 1838 to hire 'the best Leicester ram ever bred' from the famous Leicester breeder Mr Sonley, from Helmsley in North Yorkshire. Apart from being of enormous size, this ram had a striking dark blue/black head. One of his lambs, born in the spring of 1839, not only inherited his father's size and blue/black head, but, unusually, the rest of his skin was almost black and covered in fine, white, lustrous wool. He grew to be huge, over 32 stones as a two-shear, and was euphemistically described as 'extremely active', and judged to be the best ram in the north of England. His owner named him 'Bluecap' and it is from him that the Wensleydale breed descends.

Outhwaite refused 100 guineas for him in 1841 (when you could buy a small farm for 300 guineas), although he hired him out to breeders across Yorkshire to use on their Teeswater flocks. It was found that he strongly passed on his improving characteristics to his sons and he became much in demand for crossing with the local Teeswaters on both sides of the Pennines. His blood then flowed down the generations without any further outcrossing, combining the hardiness, activity, size and lean flesh of the Teeswater with the early maturity and better carcase of the Leicester. The new breed also bore Bluecap's distinctive dark coloration of the skin, particularly on the head and ears. There is nothing comparable in the sheep world to being able to trace the emergence of a breed to a single mating with a single ram. However, just like the Border Leicester (and the other Leicester descendants), the Wensleydale's true vocation only emerged decades later when a first cross with local hill ewes was found to produce a superior type of hybrid.

Breeding and showing Wensleydales is very much a minority interest and rather out of fashion. It takes an obsessive attention to detail coupled with empathy for the sheep in your charge, amounting to love. Mark Elliot has both. Early in life he determined to follow his calling to breed Wensleydales and he has never repented of his decision.

Mark is widely acknowledged to have a genius for preparing sheep for showing. He knows instinctively how to clip and shape the wool to enhance the best points and diminish the worst. He understands the perfection he is aiming for, and has the perseverance to keep trying to achieve it even though he knows he never will. He has shown his Wensleydales at most of the national shows over the years and won at most of them.

The Great Yorkshire Show is the one he values the most, not for parochial reasons, but because the competition is stiffest. To win there is to know you have reached the top.

There are many simple tricks, such as that the pirl of the wool can be improved by swimming the sheep a few days before showing, in full fleece in clean flowing water, for just the right length of time. If it was summertime, Mark's father would take his show sheep to Pateley Bridge and swim them in a pool in the River Nidd. He would strip off and swim in a deep pool with each sheep on a halter. On one of these outings, as he splashed around in his underwear, he heard his arch-rival's voice from behind some trees on the bank, 'Ah your secret's out now, Elliot!'

He also used a special pirl dip that put a spring in the ringlets, like giving them a perm. But they no longer use pirl dips, although Mark would not tell me what he has replaced them with. It's a trick of the show-ring and part of the showman's art. Dipping sheep for parasites is not routinely done now because they can be inoculated with a systemic preparation that does the same job and avoids manhandling them. This is a welcome innovation that avoids wrestling sheep the size of Wensleydales into the dipping tub.

A remarkable feature of the Longwools, in particular the Wensleydale, is that whereas the quality of most sheep's wool varies depending on which part of the body it grows – neck, leg, tail and belly wool is inferior to the rest – a Wensleydale fleece is the same all over, even on its legs and the mug on its head, and the long-staple lustre wool is pirled to the very end of each ringlet. It is the highest-priced white wool in the BWMB's schedule, which in 2011/12 was 380p a kilo – which made the average fleece worth about £25. The main market is in the manufacture of worsteds for the best suiting cloth. Also unlike short- and medium-wool fleeces, which come off as a whole, the fleece of a Longwool divides into two halves, parting along the spine. Mark Elliot clips his sheep standing up (both he and the sheep) sweeping the shears in hoops from belly to backbone, parallel with the ribs, rather than laterally across the ribs where the cutter would tend to leave unattractive lateral steps in the growing wool.

Pedigree livestock breeding induces intense rivalry between breeders, who, by definition, are individualists with strong opinions about the attributes of their breed and the rules that ought to govern their pedigree. Fierce disagreements often arise over arcane points of principle that would appear trivial to an outsider. To criticise a man's sheep can be felt as keenly as if you were criticising his wife or children, and in such a small world, where everybody knows everybody else, slights are seldom forgotten. One of these differences arose almost at the beginning, when the Wensleydale breed achieved Flock Book status in 1890. 'Flock Book status' means that it was recognised as a pure breed and therefore registration in the breed's Flock Book (its register) would be accepted as proof of the lineage of the animals there recorded and that they satisfied the breed standards as laid down by the Society.

The early breeders who tried to form the Wensleydale Longwool Sheep Breeders' Association fell out over the question of whether any ram used before 1889 (the year before the Flock Book began) was entitled to have his pedigree registered. The older, established breeders could prove the purity of their rams for upwards of twenty years and refused to accept that they could not be registered. They therefore seceded from the Association and formed the Wensleydale Blue-faced Sheepbreeders' Association which ran in parallel with the WLSBA for thirty years until a rapprochement in 1920 ended the feud. This arguably cost the breed the advantage over its great rivals the Border and Bluefaced Leicesters and created much lasting bitterness amongst its breeders.

Although the feud might have given the Border Leicester

an advantage in the wider sheep-breeding world, it did little to slow the Wensleydale's growing local hegemony over its progenitor the Teeswater, which was eventually reduced to a rump of a few breeding females. To the inexpert eye this eclipse might be surprising because there is little apparent difference between the two breeds and they both do a similar job. But a closer look will show that the Wensleydale is bigger and bluer, grows more wool and carries itself with an impressive hauteur. Teeswaters' ringlets do not cover their eyes and head as thickly as the Wensleydale's. It is claimed that a covering of wool over the head gives protection from flies and sun, but it is hard to see anything in the claim because most other breeds survive perfectly well without a mug of wool on their forehead. Comparing the two breeds it is difficult to understand why the minor differences between them should matter so much. Especially why the two breeds, so closely related, should breed true to their own type when the only genetic difference is that 200 years ago one Teeswater received an infusion of Dishley Leicester blood.

✒ 8 ✒

THE WELSH HALFBRED

I T IS FAIRLY EASY TO GET SHEEP TO BREED WITH ONE another. Rams of most breeds will mate with ewes of most others. Now and again a ram will refuse to mate with sheep from a different breed. I once bought a Bluefaced Leicester ram lamb and put him with a little mixed flock of Swaledales and Herdwicks. He served all the Swaledales but he wouldn't go near the Herdwicks. He didn't like the look of them, or maybe they didn't smell right, but he wasn't the slightest bit interested; even when the ewes made it pretty obvious what they wanted, it made no difference. He just didn't see Herdwick ewes as marriage material. I even saw him butt one out of the way so he could mount a Swaledale. But this kind of fastidious behaviour apart, sheep being sheep, most rams will breed with most ewes, given the opportunity, and produce lambs of some sort, because all sheep come from the same genetic stock.

But it is a different matter altogether to create proper hybrids. Maybe my Bluefaced Leicester ram was wiser than I thought and knew there was little point in wasting his energy impregnating Herdwicks, when he could do it with Swaledales, who were just as keen to bear his offspring and continue his line. What point was there in coupling with a

Herdwick, whose progeny would be pedestrian and disappointing, when he could create Mules and be triumphantly proud of his issue.

When hybridisation works, it produces very pleasing results, sometimes from the most unlikely pairings. And of course if it produces horrors it is always possible to do a Bakewell and eat your mistakes. The supreme example of a hybrid that works is that 'Queen of Sheep', the Scotch Halfbred. This combination of a Border Leicester and a Cheviot is the benchmark, the template, against which all other hybrids must be measured. At one time it had no rival as a ewe for crossing with meat-producing rams such as a Suffolk or Texel. But gradually over the last few decades it has found its centuries-old dominance challenged by such upstart hybrids as the Mule, with its socially ambitious Bluefaced Leicester sire. There has also been another pretender to the crown which appeared a few decades ago; for a while it seemed as if it might succeed in its challenge, but in recent decades it has slipped back.

The Welsh Halfbred on its mother's side is from the mountains of Wales, and its father is a Border Leicester from the Borders of Scotland. There are 7 million sheep in Wales adapted to every nuance of climate, terrain and their breeders' fancy. About 2 million of those are Welsh Mountains, which by 1958 were so numerous that the breed society divided itself into two sections: the Pedigree Section and the Hill Flock Section. The latter concerns itself with the Hardy Welsh Mountain, and it is from this breed that the Welsh Halfbred was created nearly sixty years ago.

Until 1955 the Welsh Mountain was outside the magic circle of hill breeds, each of which had its own complementary Longwool crossing sire. Welsh draft ewes were usually sold in the autumn into a market where the only buyers were either butchers, looking for meat to make mutton pies and curry, or farmers seeking a cheap ewe for crossing directly with a Down ram. The lambs from this cross would be good enough for the meat market, even if on the small side, but they would seldom be of the highest quality, and once the ewe had given a crop of lambs, she would usually be sent for slaughter at the same time as her lambs, at the end of their first summer in the lowlands.

There had been some haphazard and disorganised crossing of Welsh draft ewes with Border Leicester rams, but there were no recognised standards for the resultant cross-bred lambs, and no guarantee of the quality of sheep for sale. Those buyers who might have been interested were given little confidence in the quality of the animals and, as a result, Welsh hill farmers received some of the poorest prices for their draft ewes of any of the British hill breeds. This was not helped by their being small, half-starved and straight off their wild mountain grazings. But their appearance did not reflect the true hybrid-breeding potential of these sheep, or recognise that they were still young when they were sold, having bred only two or three lambs. They were destined for the abattoir almost before they had had a chance to mature.

It was not until three ambitious young farmers from North Wales saw an opportunity and determined to realise its potential that the Welsh Mountain ewe's fortunes changed dramatically. They could see that the first cross with a Border Leicester could produce a breeding animal that would rival the great 'Queen of Sheep' itself. Properly promoted, they were confident that the Welsh Halfbred would find a ready market. In 1955 the three 'young Turks', as Nick Archdale calls himself, the late Francis Morris and the late Gordon Wilyman, formed the Welsh Halfbred Sheep Association with the aim of breeding a hybrid that would produce high-quality butchers' lambs and lift the fortunes of the Welsh Mountain draft ewe from its post-war nadir.

Maybe it was because both Nick Archdale and Gordon Wilyman were outsiders and had not been brought up to stock farming, that they could see the opportunities. They were certainly hungry for the considerable success they both achieved as farmers and judges of livestock. Nick Archdale says the credit for the success of the Welsh Halfbred should go to Gordon Wilyman, who gave the Association 'the stamp of integrity'. But Nick plays down his own efforts that went into developing the Welsh Halfbred. Right from the start the Association set, and enforced, very high standards for the sheep sold under its auspices. Every seller had to be a member of the Association, although the membership fee was set at a nominal one-off £5 so that nobody was excluded on account of slender means. No sheep could be called a Welsh Halfbred unless it had been sold at one of the Association's annual Welsh regional sales. All ewe lambs had to conform to a uniform standard of appearance and the seller had to warrant that they had been sired by registered pedigree Border Leicester rams. The Association appointed inspectors who examined every single sheep in the pens before each sale, and had the right to reject any animal they judged defective without giving any reason. And every sheep was warranted by the Association to be the age the seller said it was and free from specified parasites, diseases and defects.

The officers of the Association took their duties very seriously and were determined that buyers could buy breeding sheep with complete confidence. Commercial buyers of hybrids have no knowledge of, or control over, the breeding of the ewes offered for sale. In most cases they will only find out their qualities after they have had their first crop of lambs, when it will be too late. This is one of the reasons why the Welsh Halfbred Association's guarantee was so effective in giving buyers the confidence to trust to the honesty of the breeders of the parent breeds.

The morning after one sale, Nick Archdale received a complaint from a buyer near Beccles, close to the Suffolk coast, about the state of the feet of some ewe lambs he had bought the day before. So he and Francis Morris left immediately for Beccles, driving across England, in pouring rain, long before motorways. When they arrived and turned up the sheep, they found a few with mild foot rot, which they treated with the usual gentian violet antiseptic spray and then drove back to North Wales, all in the day. That attention to detail ensured the Association's success in creating a huge following for the Welsh Halfbred, almost overnight, and almost from nothing. At the height of its popularity up to 50,000 breeding ewe lambs annually went through the sale-rings in one of the Association's five autumn sales at Ruthin, Builth Wells and Welshpool.

This threw a financial life-line to Welsh hill farmers who had hitherto struggled to make much of a living from their flocks; the better prices encouraged them to breed a stronger, better Welsh ewe, no longer seen as fit only for making lamb curry and mutton pies. Within a few years the Welsh Halfbred became the Welsh hybrid and quickly established a powerful national following. It had the strengths that come from hybridisation: its mother's hardy constitution, thriftiness, and milking and mothering instincts; and from its Border Leicester father, size, prolificacy, meaty frame, and good teeth and feet. Of course, it had the usual demerit of hybrids that they are a cul-de-sac, the end of the line. Farmers who keep them cannot breed their own replacements, and once hybrid ewes reach the end of their productive lives, they must be replaced with more hybrids.

The Welsh Halfbred is smaller than the Scotch version, but it is cheaper to keep, lives easier off the land, and will produce two lambs of similar size to their mother, each year, with minimal extra feeding. It has the robust look of the barrel-shaped Leicester, with white legs, white face and erect ears, black muzzle and a good fleece. In short, it is a great grassland breeding sheep.

Nick Archdale is a lean, tall and energetic man of

eighty-nine. His forebears emigrated to Co. Fermanagh from Shropshire in 1605. He describes himself as 'a complete hybrid': Northern Irish father, English mother who couldn't have been more English, he was born in Rhodesia, married a Welsh wife and lives in Wales. Little wonder that he appreciated the value of hybrid vigour and he should have built his farming success on it.

The Archdale sheep enterprise at their farm at Pen Bedw is a text-book example of the sheep pyramid on one farm. Right at the top is a small flock of 100 pure-bred, high-quality, pedigree Welsh Mountain ewes to breed rams as sires for the main hill flock. The best twenty ram lambs are kept each year to breed with the main hill flock of 1,000 Welsh Mountain ewes. This flock is descended from a ram lamb born to a ewe that survived the awful winter of 1947. She lived for weeks under the snow, eating heather (and even her own wool). The flock grazes the 1,800-foot hill Moel Famau (pronounced 'moyle vammer'), from whose summit, on a clear day, can be seen much of North Wales, the Isle of Man, up to the Cumbrian fells and across the Lancashire plain as far as Blackpool Tower. In 1810 the 'Grateful Farmers of North Wales' subscribed towards the building of an ambitious stone tower in the form of an Egyptian obelisk on the summit to commemorate the jubilee of 'Farmer' George III. The structure was never completed because the subscribers ran out of money, and then much of what had been built blew down in a great storm in 1862.

The pure-bred mountain flock forms the genetic core of the sheep enterprise upon which all the rest of the farm's flocks depend. The best 300 ewe lambs are retained in the flock each year to replace the same number of five-year-old ewes that are drafted out at the other end. This keeps the numbers steady at about 1,000 breeding ewes. All the surplus lambs are sold: the ewe lambs for breeding and the wether lambs for meat.

But instead of his selling the 300 older ewes off the farm

(as most hill farmers would have to), there is enough better land at a lower level at Pen Bedw to keep a flock of about 900 draft Welsh Mountain ewes, in three age groups, all of which are crossed with Border Leicester rams to produce Welsh Half-breds. This is the first crossing in the pyramid. Each year, the best 400 or so Halfbred ewe lambs join the main flock of 1,800 Welsh Halfbreds which are kept on the best land on the farm. The rest of the lambs of both sexes are sold. All these Half-breds are crossed with meat-producing rams, usually Texels, which are found to perform best at Pen Bedw. This is the second cross in the pyramid. There is no fixed retirement age for these Halfbreds. They will be kept in the flock as long as their feet and teeth are good.

This is how the sheep pyramid works on one farm. At each successive level, as the ewes mature and move downhill, not only does the lambing percentage increase, but the lambs pro-duced become bigger and meatier. The four flocks at Pen Bedw contain a total of about 4,500 breeding ewes, which produce about 6,000 lambs a year. These are made up of about 1,200 pure-bred Welsh Mountain hill lambs: 1,250 lambs from the draft Welsh ewes crossed with the Border Leicester and about 3,000 Texel cross lambs from the Welsh Halfbreds; all of these are destined for the table as high-quality Welsh lamb.

Pen Bedw is unusual because few farms are large enough or, more to the point, have such a range of suitable land from high hill to lowland to keep a flock at each level of the pyramid. For many years they bought in Welsh Mountain draft ewes, and were the biggest sellers of Welsh Halfbreds in Britain. But the trade has recently shrunk from the great phenomenon it once was. This is partly due to the Welsh Halfbred losing out to the Welsh Mule – a cross between the Hardy Welsh Mountain and the Bluefaced Leicester (that Bluefaced Leicester again!), which is believed to be more prolific, but the lambs are not as tough because they are born with very little wool, and in a cold wet spring more die after lambing than those sired by the hardier Border Leicester.

The Bluefaced Leicester has also tended to eclipse the Border Leicester because the latter's breeders have, as Nick Archdale said, 'concentrated too much on the front end rather than on the back end, where the money is'. Border Leicester breeders seem to have yielded to the temptation to produce stylish sheep with an aristocratic head, but as Bakewell might have put it, nobody eats a sheep's head. For this reason the Archdales have introduced a switch of the meaty Texel into their Border Leicester rams to try to counteract the front-end dominance of the Leicester; this coincidentally was the chief demerit of Bakewell's original Dishley Leicester and seems to have a tendency to reappear in its descendants if unchecked. This infusion precludes them from selling Halfbred ewe lambs at the official Welsh Halfbred sales, because they cannot be warranted as having pure Border Leicester fathers.

The Welsh Halfbred's great strength was its teeth and feet. These, combined with the quality of the udder, are of paramount importance to the longevity of sheep and particularly important in hybrids that are expected to work hard. A ewe has to be able to provide copious amounts of milk, in a concentrated period of time, to feed two lambs to be as big as their mother in four to five months. It is therefore crucial that she 'keeps a good hold of her bag', as they say in ovine circles. For if her udder begins to sag there is a risk of her treading on her teats, or picking up an infection that can cause mastitis, or other injury. Also her lambs will find it harder to suckle from a ewe whose udders hang low and the ewe is likely to be less agile and energetic if she has to carry around a pendulous udder.

There have been many other efforts over the years, especially in Scotland, to create hybrids, some more successful than others. The Scottish Greyface is a Border Leicester ram on a

Blackface ewe that has a strong following. The Shetland-Cheviot is a brave attempt to put some value into the white Shetland draft ewe by crossing her with a North Country Cheviot ram. And going rather against the trend of subordinating the wool to the carcase, 'breeding experts' at Scottish Fine Wool Producers bred the Lomond Halfbred. This involved the creation of a new breed from two types of Merino, which they called the Lomond and that became the crossing ram for white Shetland or Cheviot draft ewes. The Halfbred ewes clip a high-quality fleece and as lambs they can be shorn twice in their first year, as can their cross-bred lambs before slaughter.

This hybrid has not met with universal approval. It is not as good in the carcase as the other popular hybrids and the extra wool hardly makes up for that unless the price is to rise considerably. Its creators seem to have forgotten, if they ever knew it, Bakewell's famous dictum that you can't have fine wool *and* a superior carcase. Also, as the nineteenth-century breeders discovered, after a couple of generations the Merino will ruin the carcase of any sheep it breeds with. It is hardly surprising, therefore, that the Lomond Halfbred does not seem to have had quite the success its progenitors hoped for.

Until the early years of the nineteenth century if a farmer wanted to sell his sheep (or any other produce or livestock) he would take them to his nearest market town on market day and try to strike a bargain with any buyer who approached him. There were also bigger sales on fair days, but the same method applied: buyers and sellers haggled until they either made a bargain or the seller would have to take his produce home and try again next week. This was how things worked in England for many centuries.

Then, in 1849, Robinson Mitchell, a man from Cockermouth, in Cumberland, who had started a weekly furniture sale on the Fairfield, declared that he was 'tired of seeing the higgling and piggling which it required in order to make a five-pound bargain', and began to take open bids from buyers. It is claimed that this is where the modern auction system began. But this is an ambitious boast, because the Ancient Greeks sold things by auction, particularly maidens. Auctions being as old as time, Mitchell's claim that he was the inventor of the system may be going a little far, but it is true that he was one of the first to revive the system for the sale of livestock and one of the first in the country to erect a purpose-built auction mart in 1865. This was built close to the new railway station that so radically transformed the movement of livestock and killed off the ancient droving trade. Animals could be sold in the morning and transported across to the other side of the country by nightfall. This was hardly, if at all, slower than road transport today. Very soon similar auction marts grew up close to railway stations in market towns and remoter places all over the country.

It is now hard to imagine sheep farming without auction marts. They have many benefits to both seller and buyer, with the most important being that they are the quickest way to turn goods into money because the contract is made, and the price becomes due, on the fall of the hammer. They are also one of the purest markets by which buyers and sellers are brought together and a price is fixed immediately. Few people would be rash enough to dishonour an auction debt; it would be social suicide in such a small world as farming and auctioneering, where reputation is almost everything. No auctioneer would ever again take a bid from a defaulter.

Auction marts have always been more than just a place to buy and sell. They're where farmers go for their weekly get-together, and a respite from the endless farming routine, where they can catch up with the news, and have a whisky or two. Some use the auction as a kind of bank, keeping a running account, and when they sell livestock they will leave the money 'lying on' and when they buy the auction will deduct the price from their account. Auctions will lend money to favoured people. One dealer I did business with borrowed the money from an auction mart to buy fifty acres of land and paid it back over a few years.

When I worked for Harry Hardisty, the Herdwick breeder (we shall meet in Chapter 10), we used to go to the auction most weeks in winter to sell fat sheep. Before the start of every sale the auctioneer would get somebody to draw a ticket out of a hat, like a raffle, to decide which pen to start with and then he would hand over to the mart men, who were responsible for ensuring an orderly flow of animals through the ring. They would not hold up proceedings to wait for the seller if he wasn't present and they were reluctant to skip a pen because it would confuse the order of sale and if they did and the sheep made a bad price, they tended to get the blame. The sheep to be sold were penned down one side of the mart, and those that had gone through the ring were penned down the other side. The men worked their way up the aisles, driving a pen at a time into the ring until all the pens on one side were empty and all the pens on the other side were full. If sheep didn't make what the seller thought they were worth, he would 'pass them out' and take them home. Once Harry was so exasperated with the bidding for a pen of his fat sheep that he pulled off his cap, beat the air with it and, slowly shaking his head, shouted up at the auctioneer on his rostrum, 'Let them out! I'll take the buggers home and eat them myself!'

Harry wasn't much of a showman. He was too shy to cope with being the centre of attention. Occasionally this cost him a better price. Sometimes you could wring another pound or two out of the buyers by putting on a little show or making some effort to make your sheep stand out from the hundreds of others. Some sellers were naturals at it. One trick was to sort the sheep by size and colour of wool into uniform batches of ten or twenty, which always made them look more attractive.

Another was for the seller to stand in the middle of the ring, where the floor is slightly higher, and make the sheep parade round and round him like a circus ringmaster with a troupe of ponies. The knack was to wade into the middle of the flock and exploit the tension between the animals' fear of the faces arrayed around the outside of the ring and their desire to flock together for safety. The seller would stand amongst them and tap them round with his shepherd's stick. Once the first couple of sheep moved, the rest would follow, trotting nose to tail, round and round the ring, parading themselves before the buyers.

This kind of showmanship was often best seen at the autumn sales of draft mountain ewes in country auction marts at remote places, like Reeth, or Middleton-in-Teesdale in the high Pennines, or the Herdwick sales at Troutbeck or Broughton-in-Furness. The auctioneer would announce the next seller, the gate would open, and a lean, weather-beaten hill shepherd (often accompanied by his dog) would stride into the ring, in his nailed shepherd's boots and waterproof leggings, long coat flapping behind him, smoking pipe clamped between his teeth, and take up position in the middle of the ring. The first batch would be driven into the ring and he would make them parade around him, encouraging them by tapping their steaming backs with his long stick. These sheep were often straight off their mountain grazings, dipped and marked and sorted into regular batches. If there was a large 'show' it always excited interest and in the hands of an expert it was an impressive sight.

Certain cosmetic tricks could lift the bidding. Sheep that ought to have had white faces and legs would have them washed with Fairy Liquid and their wool shaped with shears – particularly round the head and neck and rear end – to try to give the impression that under the fleece lay a promising carcase. It was hard to know whether it had much effect on the price. Many of the buyers were hard-headed men who were not easily taken in. Harry thought that titivating sheep for sale had a similar effect to a woman wearing makeup: 'You know she's wearing it but you don't know what she would look like without it.' Sometimes fresh life could be breathed into flagging bidding by holding up (or sticking on the back of a bullock) some folding money – twenty pounds would usually do the trick – as extra 'luck money'. This is the cash that sweetens the deal after the sale. The seller will seek out the buyer, and offer his hand with a folded banknote in the palm, to seal the bargain, with a handshake, for 'luck'.

🐏 9 🐏

THE SUFFOLK

It profiteth the lord to have discreet shepherds, watchful and kindly, so that the sheep be not tormented by their wrath, but crop their pasture in peace and joyfulness; for it is a token of the shepherd's kindness if the sheep be not scattered abroad but browse around him in company. Let him provide himself with a good barkable dog and lie nightly with his sheep.

From a thirteenth-century treatise on estate management, quoted in Eileen Power,
The Wool Trade in English Medieval History (1941), lecture II

THE HYBRID EWE IS ONLY THE FIRST CROSS IN THE sheep pyramid. To fulfil her purpose of breeding high-quality butchers' lambs she has to be crossed with one of what are broadly called the Down breeds, which all have the same purpose of siring offspring with meaty carcases. This is the second and 'terminal' cross in the pyramid and is where we get most of our lambs destined for the modern meat market. This group of sheep includes many breeds that have never seen a piece of English downland, but they all have in common that they are either directly or indirectly descended from the Southdown, which was the first modern Down breed, created in Sussex largely by John Ellman (1753–1832) over 200 years ago.

In 1780, a decade after Bakewell's work with the New Leicester, John Ellman of Glynde in Sussex began breeding sheep to satisfy the growing demand from a metropolis becoming hungry for meat. Ellman is credited with turning the native shortwoolled heath breed of the South Downs and the other chalk hills of Kent, Sussex and Hampshire, into a meat-sheep for the London market. And in so doing he created a sheep that not only became the premier breed of the English downlands, but also had a crucial influence in the formation of all the English Down breeds which are the basis of sheep farming across the New World. For, apart from being the ancestor of the Down breeds, the Southdown was, and remains, particularly in the Antipodes, an important breed in its own right. In this it was more successful than the New Leicester, which, once it had done its work, rapidly sank into obscurity. The vital difference between the two was that the Southdown was an improver of its own heath-type relatives, which became the Down breeds, whereas the New Leicester's value was as a crossing sheep with ewes of entirely different ancestry.

The old type of Southdown that had ranged the uplands in the south of England for centuries when Ellman began his breeding improvements possessed a number of admirable qualities that few other English breeds of the time could match: their short wool was better than any of the other Shortwools, apart from the Ryeland; they were slender-boned with well-flavoured flesh; and importantly their hindquarters, where the most valuable cuts are, were heavier and stood higher than their forequarters; also they not only matured earlier than all the other unimproved British breeds, but they were very frugal users of pasture and the wethers could be ready for the butcher at eighteen months.

By the time Ellman began his work Bakewell's changes to domestic livestock breeding had already inspired a great sea-change in the type of sheep the breeders were creating.

It has been said by various modern commentators that before Bakewell breeders thought the only way to make their livestock more productive was to feed them better. It is said that the practice was to send their best animals to market and retain the poorest for breeding and that they did not understand the harm it did to their flocks. But it is hard to believe breeders did not understand that breeding from their poorest animals would cause their stock to deteriorate. It seems more likely that this assertion has been misinterpreted and that breeders sent their 'best', i.e. their fattest, animals to market and kept their 'poorest', i.e. thinnest, to breed from. It has to be conceded that in most cases there was no batching according to age, animals of all ages ran together, and when an individual was ready for slaughter it would be drawn out from the flock and taken to market, rather as they still do on North Ronaldsay.

Ellman was only one of the several improvers of the Southdown working at the time, but his gift for self-promotion, ably supported by his friend and supporter, Arthur Young, the first Secretary of the Board of Agriculture, ensured that he got most of the credit for its improvement, while Young claimed the credit for being the first to import Southdowns into Norfolk. Together with certain aristocratic enthusiasts, notably Thomas Coke of Holkham and the Duke of Bedford at Woburn, they made the Southdown the most fashionable breed of the day. It was favoured by 'gentlemen farming their own estates', not least because its mutton was reckoned to be of far finer quality than the New Leicester, which was widely considered fit only for the labouring classes. By the middle of the nineteenth century the flesh and wool of the Southdown fetched the highest price of all British breeds.

As with most of the early livestock improvers, Ellman was, if not quite secretive, at least opaque about his methods. He is said to have introduced into the old Southdown a cross of the now extinct dark-faced, polled Berkshire Nott, and, as is the inevitable result, sacrificed the fineness of the wool and sweetness of the meat in favour of size and earlier maturity. But the deterioration in the wool may also have been down to better feeding, rather than any genetic alteration. Ellman seems not to have followed Bakewell's practice of in-and-in breeding and ruthless culling of animals that did not quite come up to scratch. Rather, he mated those of his ewes with undesirable qualities with rams he had selected to correct these defects.

There seems to have been some Roman ancestry in the old Southdown, with its polled head, white Roman nose and fine wool, but there is something else there as well, because early writers describe a breed with a dark, sometimes mottled face and long legs; it is also a Shortwool, not a Longwool of the type the Romans almost certainly imported. Also the French version of the Southdown has retained the dark face and long legs. Whatever its ancestry, Ellman's improved Southdown was so popular by the end of the nineteenth century that there was not an English Down breed without an infusion of its blood; some, such as the Shropshire Down, received half their inheritance from the Southdown.

Ellman and his contemporaries' work in Sussex inspired another renowned breeder, Jonas Webb of Babraham in Cambridgeshire, to create a flock whose influence on the breed was so pervasive that, on occasion, the Southdown was called the Cambridge Down. Great rivalry arose between the Babraham breeders and the original Sussex Southdown men: the latter declared that they 'could get as good legs of mutton as Webb did but the Babraham shoulder was beyond them'. In turn, Webb's success with his rams in the show-ring and the enthusiasm of the breed's noble supporters spawned a surge of interest in the breed that led in the 1830s to the creation of a number of much-improved, and a few newly created, Down breeds that met with mixed success.

The motive for the multiplication of these breeds was

simply the continual search for more and better mutton more quickly. With beef cattle this purpose could be achieved by improving three or four breeds that would readily accommodate themselves to local conditions; but with sheep, improving the carcase and accelerating the age of maturity were made more difficult because of the influence of the climate, the soil and even the natural flora of their home pastures. Their tendency to be products of their terrain made it hard for imported sheep to adapt themselves to a change of environment. This could often only be overcome by crossing improving rams with local ewes that had become acclimatised over centuries and thus preserve their female ancestors' hereditary adaptation to the home ground. This is one of the reasons we have so many different breeds of sheep in Britain.

Gradually the new breed's dependence on a particular locality could be diminished by domestic selection, which produced strains that would adapt themselves to widely different environments. For example, over the last century, the Suffolk has evolved types that are just as at home in the Scottish Borders, the hills of Northern Ireland and Wales, or the lush lowlands. The Texel has become similarly adaptable to most locales.

The most successful of the new improved Down sheep was the Hampshire Down, created by William Humphrey from Newbury, who was so impressed with the Southdowns at the first show of the Royal Agricultural Society in 1842 that he bought a son of Jonas Webb's renowned ram Babraham to found his new breed. The Oxford Down was the largest and much in demand to breed heavyweight lambs from upland ewes. The Dorset Down was created from a cross between a polled Dorset, a Southdown and a Hampshire Down; and the Shropshire Down, from a cross between a Southdown and a local Midlands or Welsh border sheep, and for a century from about 1850 was so startlingly successful as a 'colonial ranching' sheep, and for breeding lambs for the meat trade, that by 1911 it was described as 'the most ubiquitous sheep extant' having been exported to every continent of the world. But it proved to be a shooting star, because by 1972 there were only ten registered flocks left.

The Down breed that is by far and away the most successful and has had the greatest influence on sheep farming across the world is the Suffolk, which, during the last century and a half, has been providing meat for the tables of millions of people across the world.

Until the end of the eighteenth century the indigenous sheep of East Anglia was a black-faced, horned, shortwoolled type, almost certainly of Saxon origin. It had been established for centuries on the heathlands of Norfolk, particularly on the Brecklands, the hungry sandy soils of the north-west of the county, to which it was particularly adapted. It was also naturalised in the east of England, where it bordered on the Longwools in the Fens and on the Wolds of Lincolnshire, and in Leicestershire, extending south into Cambridgeshire and Essex. It was lean, long-legged, tough and agile and well suited to ranging across the open heathlands of Norfolk and Suffolk, particularly renowned for its capacity to endure 'hard-driving'. The ewes were the best of mothers, prolific and fiercely protective of their lambs. It is likely to have come from the same stock as all the other black-faced, horned breeds such as the Swaledale, the Scotch Blackface and the Lancashire variant of the black-faced type, the Lonk. Its fleece was particularly useful; the finest part from around the neck was said to be 'equal to the best from Spain' and its wool provided the raw material for the considerable Norfolk woollen manufacture that developed around Worstead, where the eponymous worsted cloth originated.

While large parts of Norfolk remained unenclosed, the

Norfolk Horn was regarded, quite rightly, by many farmers as the only breed that could thrive on the dry heaths in summer and endure the fold–course in winter. Its main value was as a provider of fertility for the light soils of East Anglia under the traditional sheep and corn fold–course system, for without the manure and treading of these sheep the land would hardly have been worth cultivating. During the summer, the flocks grazed across the huge open heathland commons that formed the greater part of each parish, and then at Michaelmas (29 September) they were gathered up and brought onto the open arable fields, where they were folded for the winter to graze across the stubbles.

Folding is confining sheep to successive portions of the open fields with movable fences, so that as they graze across the land, their manure is spread evenly over the whole area. This kind of winter grazing (which was really just scavenging) was called *shack*. The shack period lasted until the spring ploughing – usually 25 March, the Feast of the Annunciation or Lady Day (which is the Spring Equinox and a quarter day). The owner of a fold–course had the exclusive right (which was almost an obligation) to graze his sheep over certain parts of the open fields and keep them there all winter. Folding provided the vital manure for the next corn crop – usually rye or barley because the soils were too poor to grow wheat. This was the only way the light soils could be made productive before the introduction of turnips hastened the enclosure of large parts of the heath sheep-walks and ancient open fields.

As the eighteenth-century agricultural revolution gathered pace, increasing pressure arose from progressive opinion to abandon the Norfolk breed, but the conservative, traditional Norfolk farmers resisted changing a system that had worked for centuries. Nathaniel Kent expressed their misgivings in his *General View of the Agriculture of Norfolk* (1796). Sceptical of attempts to introduce the New Leicester to the heaths of the county, he emphasised that a century earlier much of Norfolk had been bleak unproductive country, where 'full half of it was rabbit-warrens and sheep walks; the sheep [Norfolk] are as natural to the soil as the rabbits, being hardy in their nature, and of an agile construction, so as to move over a great deal of space with little labour'.

It should not be forgotten, advised Kent, that it was the Norfolk sheep that were the backbone of the fold–course system 'as they fetched their sustenance from a considerable distance, and answered penning as well as any sheep whatever'. He cautioned the improvers not to abandon the Norfolk in favour of the New Leicester or the Lincoln and to 'treat without due reverence what their forefathers and ancient custom have fashioned'. He predicted, quite rightly, that the fold-courses would fail if they were to swap the Norfolk for any new breed of sheep.

William Marshall was equally suspicious of change Writing in 1787 in his *Rural Economy of Norfolk* he acknowledged the deficiency in the Norfolk as a meat-sheep, but advocated improvement by selective breeding within the breed rather than its wholesale replacement by 'the introduction of strange breeds, nine-tenths of which would starve upon the barren sheep-walks and heaths where the native breed thrives'. He warned that it was not worth losing what had been gained over centuries of adaptation to its native soil, for 'a better chine' (the peak of the shoulder and the backbone). He suggested that if the Norfolk farmers would only pay more attention to their sheep's carcase than the colour of their faces, much could be achieved without introducing breeds from counties with a different environment. Marshall praised the breed for its prolificacy, recording that one breeder he knew got nine lambs from three Norfolk ewes in 1781, and his ewes seldom had fewer than two lambs each. And it was generally acknowledged that 'no better mutton could be put upon a table'.

But what the supporters of the old Norfolk did not foresee was that the ancient open-field farming was doomed by enclosure and the future lay not in tinkering with a communal village agriculture, but in the radical creation of large commercial farms run by business-minded farmers. The old Norfolk and the old farming, largely unchanged for a thousand years, were about to be swept away by powerful progressive forces and spirits such as Thomas Coke and Arthur Young, neither of whom had any time for the open-field farming which they thought failed to put the land to its most productive use. They were quite right that the open land of Norfolk could be more profitably cultivated if those that Young called 'the Goths and Vandals of open-field farmers were to die out and allow complete change to take place'.

So because the Norfolk stood in the way of progress, all its former virtues were turned into defects: its robust, lean, rangy frame became a tendency to be difficult to fatten; its acknowledged instinct for foraging made it a 'voracious feeder'; its independence gave it 'an unquiet disposition' and made it intractable; its long legs and agility were criticised as making it look more like a deer than a sheep and gave it 'an uncouth appearance'. It did not matter that these characteristics had been a great advantage on its native heaths and commons. Arthur Young described the breed as 'contemptible' and 'one of the usual wretched sorts found in England on poor soils', a sentiment echoed by Coke, who never missed an opportunity to vent his apparent hatred of the Norfolk breed. Annually from 1790 to 1821 Coke held his legendary Sheep Shearings, to which he invited a host of eminent guests and his tenants. He used these gatherings to parade his innovations, proclaim his faith in agricultural progress, give the audience the benefit of his opinions and, not least, to make it clear to his tenants how he expected them to farm his land.

In his speech at the 1809 Sheep Shearing, Coke reminded

his guests that 'the present institution had been formed for the chief purpose of eradicating the Norfolk breed of sheep, the most worthless race of animals that ever existed'. Coke put his tenants under considerable pressure to follow his prescription, and reminded his guests that he had 'done everything in [his] power to extirpate the Norfolk sheep ...' He proclaimed that he had

> with great difficulty induced most of his tenants to change their flock and they had found a great advantage in so doing; a few of them still retained their old prejudices, but this [he] would plainly tell them, that if they could afford to keep such an unprofitable breed of sheep upon their farms as the Norfolks were, it would fully justify [him] in raising their rents at the expiration of their leases.

At his Shearing in 1805, having given cups worth ten guineas as prizes for various categories of Leicester and South-down sheep, Coke offered a cup worth fifty guineas 'for the best ram of the Norfolk breed, provided it is deemed a good one'. He must have been pretty confident that his little joke would not backfire because in 1806 two Norfolk rams were entered, but the judges, not surprisingly, did not think either of them warranted the prize. This whole escapade does not do Coke much credit. It is not clear who the judges were in 1806, but in 1804 the great John Ellman himself and Lord Somerville (a noted Southdown devotee) had been appointed to judge the Southdowns. It is highly unlikely that any judge appointed by Coke, knowing his strong prejudice and being imbued with the spirit of the times, would have been brave or independent (or contrary) enough to see any merit in the Norfolk. Any Norfolks entered into the competition would have stood as much chance of getting a fair judging as a Russian aristocrat before a Bolshevik revolutionary tribunal.

It can't have been easy for his tenants to resist this kind of pressure. Coke was the largest landowner in Norfolk, who famously declared, when he came into his huge estate at the age of twenty-two, that he had 'the King of Denmark as [his] nearest neighbour'. His detestation of the Norfolk breed was all of a piece with his Whiggish drive for progress and profit, which went with his support for enclosure of the commons and open fields and the abolition of the fold–course system. He was right commercially – between 1776 and 1816, during his tenure at Holkham, he increased his rent roll from £2,200 to £20,000 a year – but the price was that many hundreds of thousands of country people lost forever their connection with the land.

Neither were the progressives always right in their enthusiasms. Coke's attempt to breed out the Norfolks by crossing them with the New Leicester did not succeed as he had expected, but he persevered far longer than most of his competitors. Although he had roundly denounced the Norfolk, he was 'equally convinced that the Norfolk mutton for his own table is the nicer of the two'. Coke somewhat cynically promoted the New Leicester because he believed in agricultural progress and sought profit, but although he thought its 'insipid fat meat' good enough for the labouring population, he himself would not eat it, preferring the superior lean gamey meat of the Old Norfolk.

Arthur Young claimed he was the first to import Southdowns into Norfolk, and through an accidental mating by his Southdown ram with a little flock of Norfolk ewes 'belonging to a tenant' he produced 'an entirely different breed from all the rest'. He recorded this in 1791 in Volume 15 of *Annals of Agriculture*. But it is hard to believe that Young alone should be credited with the cross-breeding that set the Norfolk on the road to its transformation into the Suffolk. Young was a shameless self-publicist and gifted writer, but a hopeless practical farmer who lost money on the three farms he rented early in his life, before he found his literary and proselytising vocation. New ideas hardly ever occur to one person in isolation and it is more likely that Young was one of many Norfolk farmers who mated their imported Southdown rams with Norfolk ewes and discovered it produced a pleasing result.

Coke's campaign notwithstanding, Norfolk farmers did not sell their Norfolk flocks wholesale. Rather they preferred the prudent course of crossing them with either a Leicester or a Southdown, until, by 1842, Professor Low wrote in his *Domestic Animals of the British Isles* that the 'perfectly pure Norfolk breed [had become] rare', something Low seems to have regretted because he said 'no finer lambs are brought to the English markets ...' than the first cross with a Southdown or a Leicester. And the remarkable vigour of that first cross could never be repeated once the pure Norfolk had gone. Gradually over the next decade or so nearly all the flocks in Norfolk and Suffolk became black-faced crosses between a Norfolk ewe and either a Southdown or a Leicester ram. The Southdown was preferred for the first cross with the Norfolk, and the Leicester ram was then mated with the progeny to produce butchers' lambs.

By the 1850s there were practically no pure Norfolk ewes left. The first cross obtained its hardihood, milkiness, prolificacy and tough constitution from the old Norfolk, its carcase quality from the Southdown, and early maturity, good feeding characteristics and heavy fleece from the second cross with the Leicester. This crossing had laid a strong foundation for the Suffolk, which by the end of the decade was well on its way to becoming a distinct breed. An effective breed society was formed in 1886 and by 1900 it had eclipsed the Southdown and by 1930 there were few Southdowns left in East Anglia.

The Suffolk Sheep Society set rigorous standards for their breed in which each sheep was given points for different attributes that could total 100 for the hypothetically perfect sheep. Feet and legs were very important – the hind legs to be 'well filled with mutton' and the back 'long and level and well-covered with meat and muscle'. The fleece was to be of 'close, fine and lustrous fibre' but, as befits a meat-sheep, it was only awarded a maximum of 10 out of the 100 points. There could be no doubt what this new breed was intended for.

Between 1900 and 1960 the Suffolk spread rapidly from its East Anglian origins across the British Isles and into the wider world and became the single most successful meat-producing breed on the planet. And despite the growing popularity of the Texel during the last few decades, it has retained its dominance, being particularly popular with breeders in Scotland and Northern Ireland. In fact the breed society headquarters is at Ballymena in Co. Antrim.

The highest price ever paid for a Suffolk sheep was in 2011 at Stirling, when a buyer from North Wales bought a ram lamb, born on 3 January 2011, for 90,000 guineas. The date of birth is significant because breeders try to have their lambs born as early as possible in the year of registration to give them the best start in life. The same buyer sold ten ram lambs at the sale for £55,000, so he didn't have to find that much to make up the difference. He described the ram as the 'most correct sheep' he had ever seen. Another ram lamb went for a big price to a Belgian scrap metal merchant who is a dedicated breeder of Suffolks and predicts a bright future for them. This is despite the modern favourites, the Texel and the Beltex, being native to the Low Countries.

Breeders have not been able to do away with the Suffolk's inclination to grow horns, inherited from its Norfolk mother. It is common for 10 per cent of lambs in pedigree flocks to have slight traces of horn and 20 per cent is not unknown, although only a tiny and diminishing proportion actually does grow horns. Most growths remain as scurs, incipient horns that never develop beyond scaly patches in the places where the horns would have grown from. Those few that form small horns are dealt with by breeders before allowing the animal to be seen in the sale- or show-ring. Horns seem to be one of those secondary genetic characteristics that are hard to eradicate without losing some other valuable or desirable characteristic such as the skin colour, prolificacy and maternal qualities they get from their Norfolk ancestor.

The inheritance of horns is closely connected with sex: large horns are dominant in the male and recessive in the female. This means that to breed a polled ram *both* his parents must be polled. But a horned ram could have inherited the gene for horns from one parent (or both) and a recessive gene for polling from the other. But a horned ewe must be pure horned on both sides; if she has inherited a polled gene from one parent she will be polled, because hornlessness is dominant in ewes. But she may have a recessive gene for horns which might only come out in her offspring.

But growing horns is a harmless inconvenience compared with the serious physiological defect that the Norfolk passed on to its descendants, and until recently had proved impossible to eradicate. This was an inherited susceptibility to scrapie. Scrapie has been recorded as a lethal disease of sheep and goats for over 300 years and doubtless has been endemic for as long as mankind has kept sheep. The disease is often recorded as devastating flocks of Norfolks on the Cambridgeshire heaths in the eighteenth century. In 1793 at Dalham, on the Suffolk/ Cambridge border, a whole flock of Norfolks either died from the disease or had to be slaughtered. It was so serious and frightening that farmers were unwilling even to talk about the disease, and if their flock was affected, they would hardly admit it.

Scrapie is one of those maladies that make us realise how little we really understand the causes of disease. Called scrapie because one of the symptoms is that affected sheep *scrape* themselves against anything they can find to scratch an apparently insatiable itch, it was not until 1982 that anybody had any scientifically based theory of just how the extreme nervous degeneration, which always ended in the affected sheep's death, was caused or transmitted. It was clear that some breeds were more susceptible than others, particularly black-faced breeds. It was unknown in North America until 1947, when the Americans blamed the importation of Suffolk sheep from Britain. There has never been a recorded case in New Zealand and only one in Australia in 1952, when four of a group of ten Suffolks from Britain showed symptoms. The whole group was rapidly slaughtered as well as every sheep that had been in contact with them, the land the sheep had grazed was ploughed up and the farm was isolated for a year. Scrapie has never been seen again in Australia. There are diseases in other ruminants with similar symptoms – BSE in cattle ('mad cow disease'), a wasting disease in deer and a degenerative nervous disorder in cats and mink. But each is specific to its own species and not believed to be transmissible across them. All these diseases are now collectively referred to as 'transmissible spongiform encephalopathy' (TSE).

Sheep do not usually show symptoms of scrapie until they are about two years old. At first they appear nervous and a little jumpier than usual, sometimes they persistently smack their lips and flick their tongue. One tell-tale symptom in Swaledales is the gradual appearance of a straight parting in the hair on the skull between the horns, sometimes extending into the wool on the back of the head. Gradually the fleece becomes dull and matted and ceases to grow. Then they begin the scraping against any solid object – gate posts or convenient rocks – and gradually become unable to feed because the itching and twitching give them no respite. Eventually, in the terminal stages, affected animals have falling fits, lose the use of their legs and eventually become paralysed and die of starvation or bloat.

Once the first symptoms are spotted the kindest thing is to have the animal slaughtered because there is no cure. None of the theories about its causes and transmission could fully explain them and, as a result, it caused a similar disquiet amongst sheep breeders as the plague or cholera did to our forebears. The impetus for research came with the 'mad cow' scare in the 1980s, when it was feared that cattle had contracted BSE by being fed offal from sheep that had scrapie and that the disease had 'jumped species' and been transmitted to humans in the form of new variant Creutzfeldt-Jakob disease (vCJD). This had similar symptoms to ordinary CJD, which had been known, since at least 1900, to affect humans. There were even reports from three French zoos that apes had been infected with BSE.

The upshot of all this was that the public was stirred into an almost hysterical frenzy by the prediction that huge numbers of people had been unwittingly infected by vCJD and the world was incubating a silent epidemic that would cause untold deaths decades into the future because of its long incubation period. BSE in cattle was thought not to show itself before the animals were thirty months old. Animal-slaughtering regulations were modified to require that no beef cattle older than thirty months could be eaten and no processed animal waste fed to farm livestock. This removed, at a stroke, the source of cheap protein that had made up a large part of concentrated animal feedingstuffs fed to poultry, pigs and dairy cattle. This deficit could only be made up by importing huge quantities of soya bean meal, mostly from South America, to supply our intensive livestock industries. We now depend on these enormous imports to maintain production at the high levels required to supply the domestic market.

The balance of scientific opinion accepts the work of

Stanley Prusiner from the University of California, on the cause and transmission of all TSEs. In 1982 he discovered tiny proteins on the cells of the nervous system, spleen and lymph nodes, much smaller than viruses, bacteria, or fungi, and entirely different from any other known organism. They appeared not to have any nucleic acid (DNA or RNA) and so they could not reproduce themselves in the same way as other organisms. He called this protein a *prion* (a contraction of 'proteinaceous infectious particle'), and gave it the code *PrP*. He found that in animals with TSE these prions had become deformed, or 'folded'. And he concluded that a deformed prion was the cause of all neuro-degenerative diseases, including scrapie.

He did not know how or why a normal prion became folded into a malign shape, or how prions replicate themselves, or how these pathogens infect their victims, or why some animals are resistant and some completely immune, even though they have prions within their bodies. For example, an affected ewe can give birth by caesarean section, yet not pass on the disease to her lamb. But if the same ewe gives birth naturally the lamb is affected. It seems the disease can also be transmitted through milk or urine, but there is no practical test to diagnose scrapie in a live animal.

The fear of silently growing forms of TSE was not allayed by our being told that prions were apparently resistant to all usual disinfectants, normal heat, ultra-violet and ionising radiation and even formalin. They could be destroyed by fire (if it was kept at 900°C for at least four hours) and heat combined with pressure for a certain period (as in a domestic pressure cooker). But for reasons not made clear, only one commercial disinfectant, 'Environ LpH', made by an American firm called Steris, was effective against them. They persisted so long in the soil that burying did not give full protection, which explains why farmers were prevented from burying dead animals on their own land.

The EU made it compulsory in 2003 for farmers to notify DEFRA if they knew or suspected any of their flock had the disease. They were then given the choice of whether to become part of a genotyping scheme, coupled with selective culling of those sheep considered to be statistically at risk of developing the disease, or have their whole flock culled. The compensation is not attractive and would not encourage a flock owner to report himself to DEFRA. There was also a *National Scrapie Plan*, a voluntary nationwide sheep-testing scheme, which is now discontinued, designed to eradicate scrapie from the national flock by voluntary culling.

However, on the basis of Prusiner's theories and research hundreds of thousands of sheep (and cattle) were slaughtered and their carcases incinerated, even though they did not actually have any disease. They were simply judged to be statistically at risk of contracting one. The risk was assessed by researchers reading the genetic codes of 1,500 sheep that had developed scrapie and comparing them with the genetic codes of 14,000 unaffected British sheep. They found fifteen genotypes that contained the putative infectious prion (PrP) and worked out the likelihood of sheep developing scrapie by reference to the number of reported cases from each genetic code. They gave each of the fifteen PrP genotypes an RCAM figure, meaning 'reported cases per annum per million sheep'.

They found that most sheep which had died from scrapie had VRQ in their genetic makeup and concluded that the greatest risk was in sheep which had that in *both* parents' genetic codes. They gave these an RCAM code of between 225 and 545 and all animals with that genetic code were slaughtered. The next most prevalent was a tenth of this at 37 RCAM and the next had an RCAM of 5. One posed such a low risk as to be negligible (0.7) and in the homozygous genotype (the same inheritance from both parents), ARR/are, there has never been any reported case in any sheep. This was the genotype of the

90,000-guinea Suffolk ram lamb sold in Stirling in 2011 and few Suffolk ram lambs with a different genotype would now stand much of a chance of making such a high price, or any price.

But things are not as straightforward as they appear. For example, the risk is not constant and varies with the age of the sheep. In some genotypes the risk peaked at two years old, while in others it was three. In some other countries where they have sheep with PrP in their genetic code, no scrapie has been reported. Nor does the research explain the existence of at least twenty different types of scrapie that have been identified in the UK. In recent years *atypical* scrapie has been distinguished from *classical* scrapie. And the overwhelming majority of sheep – between 999,455 and 999,775 out of a million even in the highest risk category – do not develop scrapie. Why not? Even lambs born to a mother with the symptoms do not necessarily develop the disease.

It became clear that to eradicate all scrapie from the national flock on the basis of statistical estimate of risk, as opposed to dealing with the disease as it appeared, it would have been necessary to slaughter a significant proportion of the nation's sheep; some breeds would not have survived at all, and many hundreds of thousands (if not millions) of healthy sheep would have been slaughtered unnecessarily. This would not only have destroyed much of what the Rare Breeds Survival Trust has been trying to do since the 1970s, and achieved the extirpation of the Norfolk, which not even the great Coke could accomplish, but it would have destroyed a large part of our ovine genetic heritage without eradicating scrapie. It is therefore not surprising that the voluntary National Scrapie Plan was abandoned by breeders. On the other hand, it has been a good thing for Suffolk breeders because it allowed them to cleanse their breed of its greatest defect and set it fair for the future.

As the Suffolk rose in the world, its maternal Norfolk line

rapidly declined into penury and obscurity. In what was nearly its last hurrah, one Norfolk ram and five ewes were exhibited by a Mr Geoffrey Buxton in the 'Extra Stock' section at the Norfolk Show in 1887. This class at the end of the sheep lines was for amusing oddities and included 'Six Egyptian Sheep bred by Prince Halim Pascha of Egypt' and 'A Two-shear Three-legged Sheep, bred by Mr Norton of Raynham'.

By 1903 the last remaining animals had become an eccentric pursuit for the gentry. The same Mr Buxton showed three ewes at the Norfolk County Show, and the *Eastern Daily Press* reported they were 'fenced in very high due to their being very active jumpers'. Prince Frederick Duleep Singh from Old Buckenham Hall showed a few, as did Russell Colman (of the mustard family). Colman's was the last flock in Norfolk, but they were swept away by 'the greatest flood ever known in Norfolk' on 26 August 1912, when over seven inches of rain fell in twenty-four hours.

Then in 1905, his grandfather having done all he could to rid Norfolk of the Norfolk, the Earl of Leicester at Holkham re-established a hobby flock of Norfolks, largely to avoid their extinction, but by 1917 this flock had also been dispersed. One Dermot McCalmont, who was probably the last breeder of the old Norfolk to keep a sizeable flock of about 100 ewes, gave a stuffed and mounted three-year-old ram to the Natural History Museum, where it is still preserved. And that was the end of the breed – or so it was feared.

But there was one remaining flock of Norfolks in Suffolk, in the hands of James D. Sayer, a farmer, cattle dealer and grazier on quite a big scale. He had a small group comprising animals he had saved from extinction by gathering together the remnants of the last few flocks in Suffolk and Cambridgeshire. He kept them at Lackford, near Bury St Edmunds, until he died in 1954, aged ninety-two. Upon his headstone, at Shropham, are inscribed the words 'His folds were full of sheep'. One of his rams and a ewe are preserved at the Norfolk Rural Life Museum at Gressenhall.

By 1947 his flock was diminished to half a dozen ewes and three rams. They were becoming hopelessly inbred. The rams had developed cryptorchidism – had only one testicle – which is a recessive characteristic that comes out through in-breeding. And the ewes had become indifferent mothers, another consequence of in-breeding. The last survivors went to the Gene Bank of the Zoological Society of London at Whipsnade Zoo, where the last Norfolk ram was mated with the last seven Norfolk ewes. It was also mated with six Suffolk ewes, to try at least to preserve some of the Norfolk blood in a back-cross – because the lambs would be more than half Norfolk through the Suffolk's ancestry. These were all transferred in 1968 to the National Agricultural Centre at Stoneleigh in Warwickshire, where the pure Norfolk ram drowned in a ditch in November 1971 before the staff had organised themselves to use the ram for breeding. When the last pure-bred female died in 1975, there remained only the cross-bred offspring from the Suffolk back-cross.

The Norfolk's extinction was not entirely in vain. It had the effect of highlighting the plight of many traditional breeds of domestic livestock and galvanising those who could see what we were about to lose to form the Rare Breeds Survival Trust in 1973. Many ancient breeds of livestock had already been lost and more were in imminent danger during the dark decades in the middle of the twentieth century when there was little interest in preserving breeds that had outlived their commercial use and few understood that the traditional breeds were a genetic resource of irreplaceable value.

Joe Henson (father of Adam Henson on BBC's *Countryfile*), Michael Rosenberg and Lawrence Alderson were founder members of the RBST and took the lead in re-creating a simulacrum of the Norfolk, the Norfolk Horn, from the cross-bred lambs out of the Suffolk. It is from these that the new Norfolk Horn is descended. This is not the same contemptible animal that Coke affected to despise, but a tamer version of the old indomitable breed.

The rams of the resurrected Norfolk Horn are quick on their feet and impressively horned, but could hardly be described as agile or fierce. Admittedly the ewes are prolific, but they resemble nothing so much as Suffolks with horns. It remains to be seen whether the breed will find a role beyond the hobby-farming niche it currently occupies. The meat sells well through local butchers, especially as the carcase is fuller now than was the old Norfolk's, due to the back-crossing with Suffolks that transports the influence of the Southdown. And, of course, they have been selected more for their carcase than either their wool or their capacity to endure hard-driving and the fold-course. There are nearly 3,000 registered animals in the Flock Book, which is impressive enough, but the increase in numbers only lifts it out of Category 1, 'critical', into Category 4, 'at risk', in the RBST's register of endangered breeds.

Many asked at the time, what was the point in re-creating the Norfolk when its raison d'être had gone more than a century earlier. If there had been a need for the breed it would have survived. Was sentimentality the motive for trying to resurrect it, especially in a form it never possessed? It is hard to give an answer to this, but it is hard not to feel that it would have been better, and rather magnificent, for it to have gone, like the doomed Highlanders after 1745, rather than be preserved, with its form separated from its function, like an exhibit in a museum.

The description from 1866 entitled 'Old Norfolk Farmer', in *Agriculture Ancient and Modern* by Samuel Copeland, serves to remind us how magnificent they must have been, those true old Norfolk sheep:

They are in their roving disposition quite equal to the heath sheep, and being longer in the legs and lighter in the carcase no common fence will stop them, and they have consequently given way to other breeds, as their native commons and heaths have been brought under cultivation. The writer has seen a flock of these sheep clear a fence with as much ease as a greyhound. The rams are sometimes very large and fierce, and are a match for any dog. At seven or eight years old they are nearly as tall as a good sized donkey, and their tremendous horns render them formidable antagonists.

〰 10 〰

THE HERDWICK

The grand secret of breeding is to suit the breed to the soil and climate.'
William Youatt, *Cattle: Their Breeds, Management and Diseases,* London 1835

LTHOUGH HYBRIDS GENERATE MUCH OF THE LAMB
we produce for the table, and about a quarter of the
national wool-clip, most of the over sixty of our
native British breeds do not fully fit into the classical sheep
pyramid. Some are specialised lowland breeds, which have
been created for a particular purpose, or have become
uniquely adapted to their terrain, and others are mountain
and upland breeds for which no satisfactory Longwool crossing sheep has been developed. One of these mountain breeds,
the one that is closest to my heart, is the Herdwick, the native
sheep of the Lakeland fells, and probably the toughest sheep
in Britain.

Herdwicks are one of the most passively subversive, semi-
domesticated animals this side of the donkey. They appear
to submit meekly to being fenced in, but their spirit is never
broken by it. If they are happy in a field they will stay there,
but it is always by consent. Their atavistic need for freedom is
never far below the surface and if they decide they want to get

out fences won't hold them for long. Like prisoners of war, they remain alert to the slightest possibility of escape, always grazing with one eye on the gate waiting for someone to leave it open – a minute or two and a few inches is all they need to make a dash for it. Turn them into a field of fresh clean grass and just watch their antics. First they will inspect the boundaries of their enclosure by doing a tour, snatching at the tastiest green tips of grass as they go, searching for weak spots. If no escape route is immediately obvious they will retreat to within dashing distance of the gate and await their chance, all the while giving a plausible impression of being engrossed in the pleasure of the sweet pickings.

I have often left Herdwick ewes, on an autumn evening, grazing conscientiously in what I believed to be a secure field, only to find next morning that the birds had flown. Sometimes there might be a few stragglers left – a forlorn remnant not agile enough to go over the top. Once they do get out they don't hang about. They haven't gone to the trouble of escaping just to linger in the next field, when the whole purpose was to make it back to the heaf on their home fell.

Although most hill and mountain sheep have a strong homing instinct, Herdwicks have one of the strongest of all. They become attached to a piece of grazing ground on the open fell and if they have to leave it their instinct is to get back there as quickly as they can. Lambs learn it from their mothers, and it is quite common to find three or four, or more, generations of sheep grazing together on the same area of mountain. On hard hills, where the sheep grew slowly, females were not allowed to breed until they were big enough, usually in their third year. On some heafs there could be a ewe with the current year's lamb and as many as five or six of her male and female offspring in regular ages from previous years.

A Herdwick can assess the weakest part of a wall at a glance. It knows where it will rush (tumble down) if it runs at it, and which bits are slightly lower than the others. That's why there is a 'breast wire', a single strand of wire, stretched tight, at just the right height near the top of nearly every drystone wall in the Lake District. It is to deflect the sheep back into the field when they try to jump over. Some clever Herdwicks use the breast wire to their advantage by running obliquely up the wall, squeezing behind the wire and bracing their backs against it with their feet on the stones, like climbers going up a rock chimney.

There are tales of draft Herdwick ewes having been sold in the autumn sales in Broughton-in-Furness – in the far south of the Lake District – only to be gathered off their home heaf at the spring gather in late March. They had travelled forty or fifty miles, to get back home across all manner of country. There is a story of a police constable on night duty in Bowness-on-Windermere coming across a small troop of Herdwick ewes scurrying north through the dim streets on their way home to their heaf on the Kirkstone Fells.

I once sold seven Herdwick gimmer hoggs to a solicitor who fancied himself as a farmer. I'd already told him his fences weren't good enough, but he wouldn't listen – being a lawyer, he knew best. I put them into his field one Monday evening and by Friday they were back in *my* field fifteen miles away. He came to collect them a few days later and assured me that his fences were now as secure as Alcatraz. 'Not even Houdini could get out!' he declared confidently as we loaded them back into his trailer. About ten days later he rang again to ask if I'd seen his hoggs. I hadn't, and neither of us ever saw them again.

The sexes are well defined. The ewes are polled and feminine, while the rams are strongly horned and decidedly masculine, often half as big again as the ewes, with a powerful chest supported by strong forelegs and standing higher at the shoulder than the rump. They should have a ruff of wool growing proud on their shoulders and nape of the neck, like a lion. They are all proud display while their ewes are busy and resourceful. The rams' horns often grow to three curls and in older sheep sometimes grow into their faces if they are not trimmed or even removed altogether. This operation is usually performed with a hand-saw either by removing a curl or by taking a sliver from the inside of the horn where it has started growing against the face. As long as it is done carefully without cutting into the quick it is little different from cutting your toe-nails. There is a limited modern fashion for breeding 'cowie', or polled, rams, because their horns can cause injury both to them and to anyone working with them. But something of their essence is lost if the horns are bred off them, and in any case horns make such marvellous handles that the benefit of them outweighs any damage they might cause. The steel-grey, coarse, weatherproof outer wool overlies a softer, denser inner coat, and their legs and faces are clear, 'hoar-frosted' white, as the breed standard describes it. Any black spotting on the face and legs is deprecated, as is 'yellow' or brown anywhere on the animal.

The usually single lamb is born in late April and May with a tight little astrakhan fleece, ranging in colour from light grey to black, with the best-bred being born jet-black with white ears. Their sturdy legs are proof of the saying that Herdwicks 'make bone'. There is a practice on hard hill farms of sending the gimmer hoggs away for their first winter to the lowlands to get the best start in life. They do not compete with the older sheep for scarce winter grazing on the fells and grow into bigger sheep if they are fed better in their first year. Herdwicks from the northern part of Lakeland were traditionally walked to the mild coastal marshes on the Solway estuary, where there is seldom a frost and snow is virtually unknown. They would spend five months from 1 November to 1 April on the marshes and then be walked home and sent back to the fell, just as the older sheep were being brought down into the in-bye fields (enclosed fields round the farm) for lambing. Many dairy farmers took in hoggs to 'top' the grass to keep it from getting straggly during the winter, and have the benefit of the 'golden hoof' without having a flock of sheep competing with their dairy herd during the grazing season.

In hard times, when grass is really scarce, Herdwicks will find a way to survive. The more enterprising will find their way onto roadside verges, jumping stone walls or getting over cattle grids by squeezing themselves close to the side-walls and tiptoeing along the edge like a ballerina; I heard of one ewe that had mastered a trick of doing a side-roll over the bars. Resourcefulness has long been a feature of the breed. Lydekker, in his 1912 account, *The Sheep and Its Cousins*, wrote that Herdwicks' 'activity is unsurpassed, as they leap like bucks and ascend stone walls in almost cat-like fashion'.

The origins of the breed are obscure. There is the usual legend that they are descended from sheep which swam ashore from a wrecked Spanish galleon that ran aground off Ravenglass in the aftermath of the Armada. The story is almost certainly a fable, particularly since there has never been anything remotely resembling a Herdwick in Spain, and the same exotic origin is attributed to other British breeds whose provenance has been forgotten. A Spanish vessel in the Armada was a man-of-war, not Noah's Ark.

William Youatt, in his classic work, *Sheep* (1840), gives a slightly different slant on their supposed maritime origin:

> In the beginning of the last century a ship was stranded on the coast of Cumberland that had on board some Scotch sheep which seem to be now unknown in that country. They were got on shore and being driven up the country were purchased by some farmers who lived at Wasdale-head, in the neighbourhood of Keswick.

Apart from Wasdale Head not being that close to Keswick, the two stories add an exotic, almost cargo-cultish aura to the Herdwick's mystique. A remnant has been cast up by the sea, as if Providence had provided them for the benefit of the flock-masters of the Cumbrian and Westmorland mountains, who must be supposed not to have kept sheep before.

The *Shorter Oxford Dictionary* says that the Herdwick 'originated on the herdwicks of the Abbey of Furness' and *herdwick* is a 'tract of land under the charge of a "herd" employed by the owner'. This comes from *wick*, which is from the Latin *vicus*, meaning village, town or farm, as in Keswick, Berwick or Greenwic(k)h. But then *wick* has another meaning from the Old English *wice*, an office or function of an official. So *wick* is the herd's jurisdiction over the flock in his care *and* the land over which his flocks graze, just as *bailiwick* is both the bailiff's office and the area over which his jurisdiction extends. *Herd-wyke* is also Old Norse for a sheep farm.

So, it would appear that any breed of sheep kept on the mountain sheep farms of the Lake District would be called a Herdwick. This might explain why the sheep in William Taylor Longmire's painting from the 1860s are described as Herdwicks even though they look nothing like today's Herdwicks.

Genetically the Herdwick is said to belong to the group of Northern European short-tailed sheep and is closely related to Norse breeds such as the Ronaldsay and the Spaelsau from the island of Froya off the west coast of Norway. But they are not short-tailed sheep and seem to have many similarities with the Hardy Welsh Mountain and other types of the great tribe of white-faced breeds related to the Cheviot. They may well be a cross between the sheep the Vikings brought over in the ninth and tenth centuries and the indigenous Celtic sheep they found in Cumbria.

Herdwicks can survive long periods of hard weather, not only without supplementary feed of any kind, but without the roughage that sustains other hill breeds, because unlike the Scottish hills or the Pennines, the thin soils of the Lake District fells are largely bare of accumulated herbage. There are accounts of Herdwicks surviving for weeks under snow, and despite being blind when they emerged, and having eaten their own wool, recovering to full health afterwards.

My affair with the Herdwick began in the 1970s, when for a few months I had nothing much to do and Harry Hardisty asked me to give him a hand as a general dogsbody on his farm called Darling How, at Lorton in Cumbria. Harry was a Herdwick man, and a dog-and-stick man – or more accurately, a dog, stick and boots man. It was a shame he died before the invention of the quad bike because it would have been perfect for him, had he been able to work one and stay on it. He didn't use a tractor because he said he didn't need one, but actually he could barely drive one. He had a car he 'kept for best', but his wife usually drove that.

His mainstay was an old Bedford van with sliding front doors, into which he could, at a pinch, squeeze about twenty sheep, held in with a 'heck' – a wooden hurdle – wedged just inside the back-doors and secured with baler twine. At every opportunity Harry got me to drive the van. We went about in all weathers with the sliding doors held open by a canvas strap which slipped over the door catch, so the dogs (and I) could jump in and out quickly. Harry sat up in front with his nailed fell boots with their turned up toes, scraping and scratching the metal floor, one hand stained red with sheep smit, grasping his horn-handled stick, and an Embassy Regal between the fingers of the other.

Smit is a Cumbrian dialect word for the fluid used to mark the sheep's fleece to indicate ownership. Red was made from haematite ore (iron oxide), which was mined in West Cumberland. Black came from graphite (plumbago), which was mined in Borrowdale from earliest times and provided the raw material for the manufacture at Keswick of Cumberland pencils. Traditional shepherd's boots, like Roman marching sandals, and infantry boots, have sprung soles, so that with the heel flat on the ground the toe is about two and a half inches off it. You walk with a rocking motion using your ankles and calves, and are propelled forward, without bending your toes; it is much less tiring than walking in flexible-soled boots. They used to be made by bootmakers in market towns up and down the land, but as rubber soles have superseded them, and country people hardly walk anywhere any more, the trade has shrunk to the last traditional shepherd's boot maker in Britain, Lennon's of Stony Middleton, in Derbyshire.

Harry and I spent much of our time touring round in the van, 'looking' the sheep, gathering up strays and generally 'keeping an eye on things'. Most of Darling How was steep fields or even steeper fellsides, and much of our looking was done through a pair of binoculars from the front seat of the van. If there were sheep to gather from land which involved any strenuous climbing, Harry would usually send a dog, or me, directing the dog with a whistle and, in my case, shouting and swearing at the top of his voice, from his vantage point on the passenger seat. Now and again the dog would do something wrong enough to cause Harry to explode in fury, leap from his seat wave his arms about and yell oaths at the top of his voice. Noise carries a long way in steep valleys, especially on still days, and at most times of the year you could be sure that there would be a few unseen walkers silently trudging up or down the fells, or through the woods. Heaven knows what they made of his outbursts of dreadful language. I often felt embarrassed about his loss of self-control, but it never seemed to bother him when I pointed out that he might have shocked some gentle walker with his profanities. To him they were 'nothing but a bloody nuisance' and probably trespassing anyway.

He had a particular phobia about 'ramblers' dogs' chasing his sheep. Whenever he came across a dog off the lead, even when it was under control and no danger to his sheep, he never failed to shout at its owner, often at the top of his voice, 'Get that bugger on a lead!' There seemed to be something about gathering hill sheep that drove hill farmers to gibbering fury, either at their dogs, anyone who happened to be in the way, or nobody in particular. One neighbour, who died young, had such a notoriously bad temper, particularly when he was gathering, and used such dreadful language, that an offended walker once complained to the village policeman, who issued him with a caution. It was not clear what offence he was supposed to have committed.

Before he turned to farming, Harry had, for twenty-five years, been the huntsman of the Melbreak Foxhounds, hunting on foot, in winter, over a sector of majestic country running west from the top of Great Gable, along the watershed between the Ennerdale and Buttermere valleys, and north along the River Derwent through Borrowdale to Keswick and Cockermouth. He knew every square yard of this country, like a fox, in his bones and in his soul.

When he took it, Darling How was a run-down fell farm with a damp, single-storey house, and dilapidated farm buildings surrounding a cobbled yard with an open midden in the middle. The few flat fields were infested with rushes because the drains and ditches had been neglected during the depression, before and after the First World War. Then the Forestry Commission acquired the farm as part of a great post-war tree-planting scheme intended to replace the timber felled during centuries of industrialisation, war and coal mining.

On first sight, the farm was not an attractive prospect. But Harry knew its potential. Apart from the cheap rent, until the Commission (as he called it) got its hands on it, Darling How had been one of the best Herdwick sheep farms in the northern Lake District. And even though the Commission had planted up most of the best parts, leaving Harry 'the bits that wouldn't even grow trees', there was still something about the terrain that made good Herdwick sheep.

From these unpromising beginnings, Harry had created a prizewinning flock of Herdwicks from the scraggy collection he had been required to take over from the previous tenant. According to the usual practice on Lakeland sheep farms, Harry bought the flock at a valuation and was expected to hand on the same number and quality of sheep to the next tenant when he left.

But there was something else that attracted Harry to Darling How. The farm had only been created when the previously unfenced Lorton common was enclosed by a private Act of Parliament in 1840. The newly enclosed land had been parcelled out amongst the erstwhile commoners in proportion to their previous rights on the common and divided up by new drystone walls. The walls were built, often on impossibly steep land, with stone quarried as close as possible above the site, and then sledged down. They never carried stone uphill. Most of these walls are still running clear across the fells, proclaiming our Victorian ancestors' faith in progress. They were functional, not intended to be beautiful and built quickly to a price, but skilled hands have produced a beautiful marriage of form and function.

As all the land at Darling How lay within the newly enclosed Lorton commons, Harry had no grazing rights on any of the surrounding huge unenclosed commons on the high fells in the adjoining parishes of Buttermere, Brackenthwaite and Braithwaite, where the ancient system of common grazing that long pre-dates the Conquest still obtained. Traditionally the heafed flocks held their ground on the huge areas of land by force of numbers but, by the late 1960s, the system had become so degraded, due to amalgamation of farms and

shortage of labour, that Harry was able to exploit the resulting vacuum with impunity. Even though he had absolutely no
right to be there, the under-grazed land was too tempting an
opportunity for him to pass up.

Those commoners who still kept flocks on the common
were not amused by Harry's antics, but there was little they
could do about it because it was not a criminal offence; their
only remedy would have been to sue him for trespass. But that
would have required all those with common rights to agree to
do it and many were not interested, because they did not exercise them. Besides which, they were all a little bit frightened
of him, because he was as tough a character as his Herdwicks.

But he had another advantage. He knew that the fencing
around the woods was worn out, with large sections missing,
so he only had to let his sheep get into the woods and they
would eventually find their way out onto the common. For
long stretches the forest fence ran along a ridge or saddle of
land 1,800 feet high. On one side were the extensive woods
and the other ran down about 1,000 feet into the valley below.
So when Harry's Herdwicks emerged onto the common they
had the advantage of being way above the commoners' flocks
that had had to work their way up from the lower slopes in the
valley on the other side. They could dominate the grazing and
during the short summer season, by sheer press of numbers,
could force the commoners' flocks to stay on the lower slopes.
Then in the autumn, having eaten out the high-lying grazings
where the commoners' flocks ought to have been during the
summer, Harry's sheep simply retreated for the winter to the
shelter of the thousands of acres of ungrazed woods. And
being forced to graze out the lower slopes meant the com-
moner's flocks were eating the store of preserved roughage
that would have seen them through the winter.

Just because Herdwicks lack a compatible Longwool to
make breeding hybrids, we should not be blinded to their

remarkable breeding potential. Every year I used to buy forty draft Herdwick ewes from a breeder whose sheep grazed the high fells between Ennerdale and Wastwater. They never saw a pickle of feed from one year end to the next and gained their whole sustenance from the alpine sedges and herbage that clung to the thin acid soils of their rain-swept terrain. When they came into the auction mart, at the beginning of October, straight off their bleak common grazings as four-year-old ewes, they were thin, grey, half-starved things, 'clapped like kippers'. But their appearance belied their tremendous promise.

Compared to their mountain heaf, the grazing I offered them must have seemed like the Fields of Elysium. No matter how bad the winter, their pride and ingrained frugality prevented them from taking any hand-feed. Then, the following spring, most would quietly, almost discreetly, give birth to twins, which by August would be as big as their mothers. The astonishing thing was that these ewes would produce twins for three or four seasons, grow half as big again as they had been when I bought them and at the end of their lives achieve a better price than I'd paid for them. Such results were exceptional because the breeder was a traditionalist who passed on the full value of his sheep to his buyer. They were what the auctioneer described as 'honest sheep.' In other words, the seller had not, before he sold his sheep, tried to get for himself some of the profit he was purporting to pass on to the buyer. He could have housed them or fed them during the winter and he would have made more for himself, but he would have deprived his buyer of some of their potential and that would have been unthinkable to him. It wasn't so much the sheep that were honest as the man who was selling them.

One of these ewes proved to me the almost unbelievable hardiness of Herdwicks. One January I found her hanging by her back legs from two strands of wire running along the top of a wall where she had got tangled up trying to jump over.

When I extricated her and set her on the ground her rear end collapsed, apparently paralysed. That didn't stop her trying to drag herself along by her front legs, with her hind quarters trailing behind. I thought she might have damaged the motor nerves in her pelvis, but as she didn't appear to be in pain I kept her to see if she would improve. I tried hand-feeding her with tempting little things that sheep like, such as locust beans, dried peas and crushed oats, but she was too proud to accept charity.

During the rest of what turned out to be a cold, wet winter and spring she dragged her hind legs around her small field, fending for herself. Even when the grass was frozen, she refused to touch the fragrant meadow hay I put out for her. Then in late April she gave birth to Suffolk cross triplets. They were small but healthy so I let her keep two and fostered the third onto a ewe that had lost her lamb.

After her lambs were weaned in July I moved her into another field, intending to send her to the knackerman when she had fattened up a bit. But she seemed to do so well and had such spirit that I didn't have the heart to get rid of her. She survived another winter, again stubbornly refusing all handouts, living on what she could forage (including the bark of ash branches that had been blown off in storms), but she never moved beyond the couple of acres in the field she had chosen as her heaf.

Then one morning in late May, after most of the other ewes had lambed, I happened to glance over the wall and there she had two small, healthy lambs. I had no idea how she had contrived to be in lamb. She reared both lambs herself because I didn't have a foster mother that late in the season. This would have been her sixth crop of lambs. The following spring she gave birth to two more lambs – this time Cheviot crosses – and reared them herself. Again I had no idea how she had managed to get in lamb. I never saw any ram with her and she certainly could not have got out of the field. The ram must have found her. She reared those lambs as well, but during that autumn she started to show signs of failing, so, with a heavy heart, I shot her. I felt it was better than allowing her to suffer, or submitting her to the trauma of being taken away in the back of the knackerman's charnel-wagon. I buried her in the corner of the field she had chosen as her last home.

There is a unique tradition amongst Herdwick breeders to *rudd* both the tups and the ewes at breeding time and for autumn show and sale. Rudding is applying a deep red-ochre paste made from iron oxide powder mixed with oil and rubbed into the sheep's back and nape. This is an ancient practice without any apparent practical utility, and seems to be a relic of almost mystical significance from an older world. It is usually only done at breeding time and, as far as I know, at no other time of year, except perhaps if a special sheep was to be sold or shown in some kind of competition.

The practice looks very like a kind of *sympathetic magic*. Sir James Frazer mentions in *The Golden Bough* the primitive belief in the magical healing power of the colour red; that among primitive peoples red is an elemental colour associated with many forms of protective magic related to energy, fertility and medicine. It is not only symbolic because in traditional medicine red is believed to have its own energising power. For example, to this day in certain Eastern countries red clothes are worn during an illness because it is believed that red drives away depression and promotes healing. Military hospitals made up the beds of wounded soldiers with red blankets. In the Balkans it is still common to see red tassels hanging from horses' head-harnesses. I remember an old aunt wearing red flannel next to her skin to relieve aches and pains. I don't know whether people still use it, but it was certainly believed to be efficacious well into the last century.

Rudding at breeding time may well be an attempt to draw

on this ancient magic, to promote fertility, to protect and purify the animal for the breeding season and to encourage good health and good luck. It 'just looks right' because it has these purposes, but it undoubtedly adds beauty by contrast to the steel-grey fleece, bringing good luck and gaiety to the autumn breeding season and lifting the spirits as the days shorten and the world slides into winter. Harry would have said I was talking fanciful nonsense, but he had no better explanation for the ancient practice.

I hadn't so much as touched a Herdwick for over twenty years. So during the slack days that come after New Year I travelled up Borrowdale to see Stanley Jackson, at Nook Farm, Rosthwaite, to find out how the Herdwick was faring in the modern world.

As I crossed the county boundary a rainbow appeared against the looming pewter sky over the Solway, which I took as a favourable omen, and bowled along into the gloom. The North Cumberland Plain stretched out towards the Solway Firth in a fluorescent panorama beneath a flat line of low grey cloud. Then the rain began. By Keswick it was sheeting down and at Rosthwaite the road was awash. Derwentwater had flooded out across the low-lying fields and brown spumy water lapped at the roadside wall. The mist was down to about 500 feet, hanging like smoke in the freezing air, with no wind to disperse it. Icy rain bounced off the tarmac and sluiced down the grey slate roofs, overflowing the down-pipes and cascading across the cobbled yard at Nook Farm. We dodged and splashed through puddles and sprinted for the warmth of the farm kitchen.

Borrowdale was living up to its reputation as the wettest place in the wettest part of England. Stanley Jackson gazed out from the warmth of his kitchen at the relentless rain and flooded fields.

'Last year was bad up here. A bloody wet summer. We had eighteen months of winter. And now look at this!' He waved the back of his hand at the curtains of rain moving slowly down the valley.

'No other breed could stand this wet. They'd die in the winter. Did you know a Herdwick ewe can absorb her foetus to survive a really hard time?'

Most of the ewes had gone back to the high fells after Christmas and the fields were almost empty and bare of grass. A few older ewes stood hog-backed, enduring the metallic rain, their white ears down and rain-washed hooves bunched-up together under them while water dripped off their steel-grey wool. On the way up Borrowdale I had noticed three ewes close to the iron railing fence on the roadside, motionless and chewing dolefully. They stared at me with elliptical almost reptilian pupils in saffron irises, pausing their chewing just long enough to conclude that I wasn't interesting, before taking up the steady grinding rhythm again, water dribbling off their eyelashes and trickling down their ivory-white faces. My heart went out to these creatures enduring the bitter January rain in that bare field.

Stanley Jackson is a tenant of the National Trust, now the major landowner in the Lake District and the owner of over 25,000 Herdwick sheep heafed to the fells and let to their tenants as 'landlord's flocks' with their farms. Stanley's farm is one of the ninety-one the Trust owns in Cumbria, greatly expanded from the original fourteen, which Beatrix Potter bequeathed them in her will in 1943, bought largely with the proceeds of her *Peter Rabbit* books. She wanted to preserve the Herdwick sheep and the unique way of life of the fell farmers who kept them, and stipulated that the farms were to be let at moderate rents to suitable tenants.

But the Trust has a difficult trust to keep – it's a delicate balance between its largely urban supporters and its farming tenants who keep the Herdwicks and understand their spirit. In a sense, the Trust is the modern secular equivalent of the great mediaeval religious houses whose lands were held in mortmain. They paid no feudal dues because no living hand owned their property and as it never changed hands its ownership was protected from the normal demands the Crown made on all other occupiers of land. In so far as was possible in feudal times, their ownership was absolute. By modern popular consent the National Trust enjoys similar protection from the state's exactions: its property is held in perpetual trust; it is prevented by statute from selling or mortgaging it; it enjoys all the fiscal benefits of charitable status; and because it cannot die it pays no inheritance tax. There are other landowning charities that enjoy similar immunities – the RSPB, for instance, which owns or controls 128,000 acres in England – but none is as powerful or as well protected as the Trust.

Stanley has an eccentric way of judging the profitability of his sheep flock by comparing what he gets for a draft ewe with the number of fence posts that sum would buy. In 1980 it was sixty posts. In 2006 it was reduced to fifteen posts. Now that sheep prices have improved a little in the last few years, it's about twenty posts. Although he's hardly making a fortune, Stanley takes a long view and will carry on keeping his Herdwicks, for despite all the difficulties he loves this vocation that gives meaning to his life.

Despite valiant efforts by various groups to promote it for its hardwearing properties, the market for Herdwick wool almost disappeared in the last decade, even though it has recovered recently in line with the general increase in wool prices world-wide. There are two grades: Light Herdwick makes a penny or two more a kilo than Dark Herdwick, but both are in the lowest grade of the Wool Marketing Board's price schedule. Some niche manufacturers are making imaginative things like durable luggage from it, but the appeal is limited. At one time it was used to make Hodden Grey cloth, an almost everlasting, waterproof rough tweed, woven in Caldbeck, later

Cockermouth by John Woodcock Graves, the man who wrote 'D'ye ken John Peel'. Peel's 'coat so grey' was made from it. But it's too coarse for most modern sensibilities.

Herdwick lamb, by contrast, is one of the less well-known gastronomic delights of England. The Queen has served it at official banquets. But until mutton fell out of favour and wool became virtually worthless, the best Herdwick meat came from wether sheep. Traditionally managed Herdwick flocks had always contained a large proportion of wethers, castrated males of varying ages, which only came down from the high fells once a year to be clipped and dipped. They cost virtually nothing to keep and would often be aged three or four or more years when they were sold. Unlike the ewes, they were unencumbered by lambs, and not compromised by motherhood and became strongly heafed to their mountain grazings, where they formed a united group that would 'hold the fell' against all-comers that might try to muscle in on their heaf. They were almost wild animals and their slow-growing carcase improved with age, especially after the wild flora of the high fells had imbued its dark meat with a distinctive gamey flavour, similar to but less heavy than venison.

The Prince of Wales has been a great promoter of Herdwick mutton. He founded the Mutton Renaissance Campaign in 2004 to bring mutton back into our diet and try to improve the income of Herdwick sheep farmers. Some of the more enterprising now sell their Herdwick meat direct to the consumer, helped by the internet, and some have developed sales at Borough Market in London, and Westmorland Services on the M6. But, as ever, producers are up against the buying power of the supermarkets, who will not stock mutton because there is a limited demand for it, which remains limited because they will not stock it.

In mid-afternoon, on my way down Borrowdale, the cloud cleared and it grew cold. A weak mid-winter sun appeared briefly before slipping down behind the high fells. It was shaping up to be a frosty night. Any creature needs special qualities of endurance to survive a day-long drenching followed by a freezing night. But when that creature's only sustenance comes from poor vegetation scavenged from thin acid soils, its resilience has to be very special indeed and proves the Herdwick breeder's boast that his sheep 'live on fresh air, clean water and good views'. They bear privation with such dignity that their cleaving to this austere landscape has almost become a virtue.

ᕝ 11 ᕝ

THE DORSET HORN AND
THE LLANWENOG

It has been said with truth that each locality tends to develop the livestock most suitable to it, and the chalk hills of Dorsetshire are probably responsible for its evolution. … It is true that enthusiasts, from time to time, are able to take a breed out of its native surroundings and do well with it, but they are no more that the exceptions which prove the rule.

From an article in *Country Life*, 7 December 1907

THE BREEDERS WHO BROUGHT THE ANCIENT TAN-faced sheep of south-west England from the medieval world into the modern age of Bakewell were not impeded by sentimental attachment to the old breeds that had outlived their usefulness. Their gaze was fixed on the future and the huge changes they wrought over the ensuing 200 years so altered the native Dorsetshire sheep that it is hard to see – and they do not care to be reminded – that the Dorset Horn they created is descended (on its mother's side) from a type that would have been ubiquitous across south-west England since the Iron Age, and almost indistinguishable from the primitive Portland. The Dorset is a breed that has grown so perfectly to fit its terrain and fulfil its purpose that it has no need of a crossing sire because no out-cross could possibly improve it.

The original old Dorsets were small thrifty sheep, with longer legs than the Portland, and characteristically black lips and nostrils. The ewes seldom needed help at lambing time because the lambs had small heads at birth, and also because they were such resilient self-sufficient sheep. These qualities were carried through into the improved Dorset Horn and were a large part of the attraction of Dorsets to the owners of sheep stations in Australia and New Zealand, where they are expected to get on with lambing on their own, without cosseting and constant supervision. They are naturally early-maturing sheep with a noble temperament.

One clue to the Dorset Horn's ancestry is that it has the unusual capacity to breed all year round – an attribute that early commentators believed to come from a late-eighteenth-century cross with the Merino. But this is not true. There are numerous accounts of Dorsets lambing out of season long before their fleeting flirtation with Merinos. Edward Lisle wrote in 1756 about one of his tenants, 'Farmer Stephens', who at Lady Day (25 March) in 1707 sold the lambs fat off some Dorset Horn ewes and thinking 'his ewes to be mutton, for they looked big', at the beginning of June sent them to the butcher, who found they were about to lamb. They must have taken the ram nearly three months before their lambs were weaned. It is almost unbelievable that both Farmer Stephens *and* his landlord were surprised at one of the defining charac-teristics of the Dorset breed. And even harder to believe that Farmer Stephens could not distinguish, at a glance, a pregnant ewe from a fat one or that he did not know his ewes would take the ram and lamb at almost any season of the year.

25 March is the Feast of the Annunciation of the Blessed Virgin Mary and the Spring Equinox, and until the adoption of the Gregorian calendar in England in 1752 this was the first

day of the year. There is a vestige of this in the tax year starting on 6 April, which is Lady Day adjusted for the days lost by
the calendar change. This is a significant day in the farming
year, when many tenancies began and ended. The years AD are
counted from 25 March in the year preceding the birth of Christ
because that is reckoned to be the date of his conception.

You can see what a Dorset Horn looks like whenever you
pass a Young's pub because the brewery adopted the breed as
their mascot and have had it on their pub signs since they built
their brewery on the Ram Field at Wandsworth, where it is said
the villagers formerly kept a communal ram.

Dorset has been a sheep county for many centuries. Daniel
Defoe in 1747, in his *Tour thro' the whole Island of Great Britain*,
describes how the downs come up to Dorchester on every side:

> even to the very streets' end; and here it was that they
> told me, that there were 600 thousand sheep fed on
> the downs, within six miles of the town ... the grass
> or herbage of these downs is full of the sweetest, and
> the most aromatic plants, such as nourish the sheep to
> a strange degree, and the sheep's dung again nourishes
> that herbage to a strange degree; so that the valleys are
> rendered extremely fruitful, by the washing of the water
> in hasty showers from off these hills.

There cannot have been half a million sheep within a six-
mile radius of Dorchester in the middle of the eighteenth
century. This is an area of 113 square miles, or 72,500 acres,
which means there would have been more than eight sheep
to the acre, which even today, with modern fertilisers and
imported feedingstuffs, would be hard to maintain. In 1840
Youatt gave an estimate of the number of sheep in the *whole
county* to be rather more than 632,000, which seems more likely,
even though it works out at two sheep to every acre. Whatever

the true number, it is clear that Dorsetshire has always sustained a lot of Dorset sheep, a breed 'peculiar to itself' being 'chiefly collected within a circle extending ten or twelve miles from Dorchester'.

The Dorset has largely been created by selection from within the breed, not out-crossing. There have been experiments with crossing with various breeds but all gave indifferent results. For example, at the height of the Dishley Leicester fascination, a cross was tried to improve the Dorset's carcase, but the results were disappointing. Then, at the end of the eighteenth century when the Merino craze was at its height, it too was tried, to improve the wool. Merinos bore (and still do) fleeces of the finest wool in the world. It was worth ten times as much as coarser English wool. And unlike most other fine-woolled sheep, for example the Southdown, it was of the same excellence all over the body, so close and fine on the back that it resisted the pressure of a flat hand. Merinos are unique sheep with a very long ancestry. They could originally have come into Spain with the Phoenicians when they arrived from the Levant, and having been improved by the Romans, they were brought to the pitch of woolly perfection under the medieval Spanish Mesta, where they were called *Transhumante* because they spent their winters in the Spanish lowlands and summers in the cooler hills of Castile. They were distinguished from the other principal breed in Spain, the *Estante*, so called because they stayed put and were not moved with the seasons, like their peripatetic cousins.

The Mesta (Honrado Concejo de la Mesta) was a powerful association of sheep ranchers in the medieval Kingdom of Castile. Until the expulsion of the Moors from the south of Spain resulted in the *Reconquista* at the end of the fifteenth century, there was a buffer zone about sixty miles wide between the Christian north and the Moorish-controlled south. Too insecure for settled farming, it could only be exploited by transhumant sheep grazing, where vast flocks grazed, regulated by the Mesta, which was the first agricultural union in medieval Europe. Its members grew rich from breeding Merinos and exporting their wool, at the time the finest in the world and, as in England, Spain's greatest single resource. The Mesta enjoyed many privileges under the protection of the Castilian crown. For example, the network of 78,000 miles of *cañadas* (ancient drove roads believed to date from prehistoric times) was (and still is) protected for all time from obstruction. These had to be maintained at 100 yards wide and the most important were designated *cañadas reales* ('royal *cañadas*'), specifically created and protected by the crown. Even some streets in Madrid are still part of the *cañada* system, and those with the right still assert it by driving sheep through the modern city. Every autumn since 1994 the Spanish Ministry of Agriculture has encouraged the fiesta of transhumance when 2,000 sheep are driven through the centre of Madrid. Upon payment of 25 *maravedis* (an eleventh-century coin) to the city council, shepherds and drovers have the right to stop the traffic and take their flocks through the streets following the routes of the *cañadas*.

So highly prized and valuable to the Spanish crown was Merino wool that for centuries the export of live sheep was prohibited on pain of death. Gradually, over time it became practically impossible to enforce the ban and in any case the king himself undermined it by giving his cousin, the Elector of Saxony, 300 Merinos in 1765 (from which the Saxon strain is founded); then he allowed another cousin, Louis XVI, to buy 350 sheep to start a flock on his model farm at Rambouillet (which founded the French strain). And in 1789 the Spanish Crown allowed further stock to be exported to South Africa, from where they were taken to Australia as the basis of the vast wool flocks of the New World. But as good as the Merino undoubtedly was, it could still be bettered by a drop of

English blood, because in the twelfth and thirteenth centuries it is recorded that Spanish breeders imported English sheep to improve the local woolly breed that became the modern Merino.

'Farmer' George III and certain of his enthusiastic landed supporters came to believe that our English wool-producing flocks could be improved in their fleece by a cross with the Merino. Apart from his patriotic wish to give English sheep better wool, the king's intention was to try to keep British weavers employed by seeing off foreign threats to our textile manufacture. Sir Joseph Banks managed to procure a couple of Merinos via France in 1785, and then the king himself got hold of some indifferent specimens in 1789 by having them smuggled via Portugal to England, where he installed them on his model farms at Kew and Richmond. Five years later he approached the king of Spain directly and persuaded him to allow the export of a legitimate consignment of his prized Negretti type to add to the royal flocks. This English importation began a fifty-year experiment in crossing the Merino with many of our native breeds which, in theory, ought to have improved their wool, but in practice was an almost complete failure. Its enduring legacy, for better or worse – and largely for worse – is that traces of Merino blood flow through the veins of most of our native breeds.

The effect on the Dorset (and almost every other breed it was crossed with, except probably the North Country Cheviot) was to improve the wool in the first cross at the cost of an almost immediate deterioration in the carcase. This did not endear the Merino to commercial English breeders who were turning away from pure wool production and striving to breed meat-sheep. The breed was also disliked on aesthetic grounds because its extravagant dewlap and folds of woolly skin were offensive to the English taste for understatement and simplicity. To add insult to injury, after the first cross, the wool deteriorated unless the ewes were back-crossed with a Merino, in which case the carcase deteriorated even further. The long Spanish concentration on excellence in the wool had created an animal that was more like a late-maturing woolly goat than a sheep and perhaps more importantly, apart from the bad carcase, the breed simply failed to adapt to English conditions.

Yet, despite all the evidence that the experiment had failed, it took decades for its most zealous partisans to be disabused of their error. Even when the truth walked before their eyes, and was plainly stated by practical breeders – even the great Coke himself – many still clung to the orthodoxy that four Merino crosses ought to produce an English sheep with Spanish-quality wool. Coke spoke, with untypical understatement, for most practical breeders, when he said at his Shearing in 1811 (the year the Merino Society was founded), 'I have reason to believe that however one cross may answer, a further progress will not prove advantageous.'

There is a plausible theory that the Merino could not adapt to its British situation because most of our soils are copper-deficient and their metabolism needed more of that mineral than our home-bred English sheep. As this theory came from a Frenchman, it ought not to be considered entirely free of xenophobia, even *Schadenfreude*, because the French succeeded not only in improving some of their native breeds with Merino crosses, but in creating an entirely new and valuable breed from Louis XVI's importation. This came to be called the Rambouillet and had a huge influence on the development of the Merino in Australian sheep farming.

The failed experiment with the Merino encouraged the leading Dorset breeders to stick to their own breed and improve it from within with selected improving sires; this they saw as a more certain (albeit slower) method of improvement. So apart from an infusion of blood from the larger Somerset variant of the Dorset to increase its size (from which it is supposed to have got its pink nose), all the breeding improvements came from within the breed itself. This tended to strengthen the Dorset's connection with its surroundings and make it even more a product of the Dorset soil.

Dorset wool was good, but not the finest, because even in the seventeenth century, when wool was still valuable, high-quality lamb, 'house-lamb', at Christmas was a more profitable market to exploit and traditional Dorset breeders had cultivated it to the pitch of perfection. Long before the eighteenth-century breeding revolution from wool to meat, Dorset farmers had developed this system of rearing prime milk-fed lamb for the London market, exploiting the Dorset's unusual attribute of out-of-season lambing.

House-lamb derives from 'housed' rather than having anything to do with domestic dwellings. Well managed, the ewes would take the ram in early April (when most other breeds are lambing) and would lamb five months later in September and October – and usually have twins. The aim was to give the lambs as much milk as possible for the two to three months leading up to Christmas, so that they would sell as a luxury food, at a high price, as finished suckling lamb, at a time of year when young lamb was otherwise unobtainable. It was a labour-intensive system practised on a large scale, and would nowadays be entitled to an *appellation contrôlée* like poulet de Bresse, Parma ham or Stilton cheese.

The ewes gave birth outside and immediately afterwards they and their young lambs were housed in barns that had been scrupulously cleaned during the summer to minimise disease. The lambs stayed inside during the night with their mothers. Then in the morning the ewes were allowed out to graze the best pastures, supplemented with good green hay, swedes, grains, clover – anything of the best that was available. At mid-morning any ewes that had either lost their lambs or whose lambs had already been sold but were still in full-milk were brought in as wet-nurses and the lambs suckled from them until their udders were dry. At midday the natural mothers were brought back in from the fields to suckle for an hour or two. Then at tea-time the wet-nurses were brought in again for an hour or so. Then at eight o'clock their natural mothers came in again for the night shift. Everything possible was done to ensure the ewes were kept in full-milk because once a ewe's milk began to decline, no amount of feeding would revive her lactation; she would just grow fat. The lambs were pampered, given good wheat straw to nibble (to stimulate their digestion) and pieces of chalk to lick (for minerals), would grow quickly on the surfeit of milk and be ready for market at about eight to ten weeks old.

Another benefit from Dorset sheep is that their manure maintains the fertility of the light, flinty Dorsetshire soils. They do well on a diet of roots and green crops and, being docile sheep, readily accept close confinement during the winter in folds. Docile as they are, shearing gangs aren't fond of Dorsets, because they're 'so uneasy, and they won't sit on their arses. Chilver lambs are the worst,' says Francis Fooks. Chilver is an old English word for a gimmer lamb – i.e. not yet a year old and has not been shorn. The corresponding phrase in the north of England and Scotland is gimmer hogg. Francis Fooks, with his brothers, farms 800 acres at North Poorton, near Bridport. Their great-grandfather established the flock in 1906.

Dorsets are 'sheep for the modern world', says Francis. It is one of the few breeds whose ewes will lamb at nearly any season of the year and the pure-bred closed flock makes fine finished lambs from grass and roots earlier than nearly any other breed, without invoking the hybrid vigour that many other pure breeds seem to need to fulfil their potential. The great value of keeping a 'closed' flock and only importing rams is that the ewes become accustomed to their terrain, develop immunity to certain parasites and diseases and are a known quantity, having been selected over generations by their breeder. So long as they are as good as Dorsets, there is no need to consider breeding hybrids.

In 1937 an 'accidental' mating between a Dorset Horn ewe and a Corriedale ram (half Merino, half Lincoln Long-wool bred in New Zealand) produced a ewe lamb without horns. When this lamb was mated back with a Dorset Horn ram the resulting ram lamb was polled. This started a twenty-year breeding experiment aimed at breeding a Dorset with all the Dorset Horn qualities but without the horns. The Polled Dorset is now a separate breed which has become more popular than the horned variety. It does not catch its horns in netting, can be packed tighter in the sheep pens, takes up less space at feeding troughs and is popular with the abattoirs that do not like horns. It is questionable whether these benefits are worth the loss of handles. But there is something else. I asked Francis why he still keeps the horned variety if polled ones are so much easier to manage and he replied, 'Horned sheep remind me of what the polled sheep *ought* to look like.'

By this he means that the Polled Dorset is not quite a Dorset Horn without horns. If it's bred pure for long enough it either reverts to being like one of the other breeds that make up its ancestry or it begins to grow horns. As with the Suffolk, many supposedly polled Dorsets grow scurs and, it seems, sheep with a better carcase tend towards being horned. Horns in the Dorset are another of those secondary breeding characteristics it appears cannot be bred out without losing some of the essence of the breed.

Until artificial fertilisers became cheaper in the middle of the last century, it was common for arable farmers to allow sheep farmers to grow root crops on their land to feed their sheep during the winter, simply in return for the manure.

But modern grass varieties that stay green and keep growing throughout the winter have tended to replace roots. They do not give as much goodness to the soil, but they are much less work and consequently cheaper than a crop of the swedes and kale which the Fooks' sheep feed on through the winter. Crops from the Brassica family are a wonderful store of energy and fresh food during the winter, and greatly increase the productivity of livestock farms. They can be grown either for their starchy roots and leafy tops (swedes and turnips) or just for their green leaves (kale, rape or cabbage). Although swedes are over 90 per cent water, it is, as the saying goes, 'bloody good water' and sheep folded on roots generally do not need extra drinking water, making folding so much easier, because there is no need to carry water to them, or let them out daily to drink. Francis has found that if you give sheep water when they're first folded they drink deeply from it once and then never go near it again for weeks on end. Drinking too much also makes their dung loose and in a wet season soils their fleeces.

On the high land in this part of Dorset, the sandy topsoil is only four to six inches deep and careful shallow ploughing is necessary to avoid turning up the less fertile and intractable yellow and grey clay subsoil. In some places the limestone bedrock pokes through the surface. The soil has a warm orange ochre tone and imparts a tint to the fleeces of the sheep folded on it. It is what Francis describes as 'hungry gutless sand'. When the sun comes out after rain, the wet flints on the bare earth shine like a pavement and without the dung of livestock it would be very unproductive. In winter it is sticky and muddy and then it dries out quickly in spring. Ideally, to be farmable, it needs 'a storm of rain every day and a storm of shit on Sundays', says Francis.

Compared with the Dorset's ancient lineage, the Llanwenog would seem to be a parvenu. With its spindly legs, no horns and a woolly topknot, it resembles nothing so much as a badly bred Suffolk in a barrister's wig.

The heartland of the breed is in the Teifi valley around Lampeter in Ceredigion in West Wales, amongst small farms in soft green dairying country with a mild Atlantic climate, where the grass comes early and grows lushly late into the year. In most winters there is no snow and little frost, although fierce winter storms race in from the North Atlantic between Cape Clear and Land's End.

Already by the middle of the eighteenth century the farmers here had established a lucrative trade in butter with the London market, which they transported by sea from ports in West Wales. This connection with the capital drew people from the Teifi Valley to seek their fortune in the 'milk business' in London, particularly in Bethnal Green, where they kept milch cows to provide fresh milk for the metropolis. Welsh emigrants could make enough money from the London milk trade to set themselves up on a small farm when they retired back to West Wales.

The area between Lampeter and Llandysul was nicknamed the 'Black Spot' (*Y Smotyn Du*) by the dominant Welsh Calvinist Methodists because its inhabitants espoused Unitarianism in defiance of the almost complete Methodist hegemony in Wales. When this group of dissentient parishes in mid-Ceredigion was dotted on the map of Wales, to the fury of the Methodists, it appeared as a black cluster of intransigence. Unitarians were considered dangerous radicals and free-thinkers, especially as they were early and vocal supporters of the French Revolution and active in the Rebecca riots in the late 1830s. Almost all the original Llanwenog breeders were Unitarians (many still are), who stuck together in their dissentient group, intermarried, lent each other money and supported one another; and out of this emerged their breed, the Llanwenog.

On its mother's side it is most likely descended from the

unrecorded and forgotten primitive Cardy, an ill-shaped little animal that was once native to the hills of Cardiganshire, Carmarthenshire and Brecknockshire. It seems to have been a member of the great tribe of ancient English Heath sheep and, as Robert Trow-Smith remarks, an exhibit in 'the ovine open-air museum of the Welsh hills' with a lineage on its native soil stretching back into prehistory.

These sheep were kept like crofters' sheep, for their wool and manure. They were not as important to the farmers of West Wales as their principal export, the little hardy black cattle, which for centuries they had consigned in droves to the fattening pastures and markets of England. But, in the late eighteenth century, the cattle yielded primacy to sheep because of the increasing urban demand for mutton and tallow for candles and a home-grown demand for wool to supply the Welsh weaving industry. Huge flocks were entrusted to drovers who walked them into the Midlands and on to London along the ancient drove roads through South Wales into Gloucestershire, Herefordshire and beyond.

In 1867 the first railway to penetrate West Wales opened from Shrewsbury to Carmarthen (via Aberystwyth and Lampeter) and the old practice changed almost overnight. For the first time in history, farmers could carry their sheep to the English market in less than a day; they could buy sheep in the morning and almost have them back home on their farms that evening. By one of those happy coincidences, about the same time as the railway opened, the Shropshire Down was enjoying its heyday as the most popular Down breed. If the railway had followed a southern route from England the story would almost certainly have been different. But as the line went to Shropshire, the breeders in West Wales found it easy to take home the Shropshire to improve their native flocks.

The Shropshire Down had been created by crossing the New Leicester with one of the ancient West Midland or Welsh border types and then by a second cross with the Southdown. The new breed first appeared in the area around Bridgnorth, where the now extinct ancient Morfe was indigenous, so it might be that on its maternal side it is descended from the Morfe, but as no trace of that breed has survived, it is impossible to know. The Shropshire was widely admired and became very fashionable. It was prolific, with a very good fleece, and thrifty, gaining most of its sustenance from grass; its only arguable defect was that it tended to sire small lambs when crossed with small hill ewes. But breeders were compensated for the smallness of their lambs by the fact that they were regularly rearing between one and three-quarters and two lambs from every Shropshire ewe.

Within ten years of the railway arriving in this remote, almost entirely Welsh-speaking part of Wales, the Shropshire Down became established in the mid-Teifi valley, largely through rams imported by three improving farmers whose land lay alongside the new line between Lampeter, Llanbydder and Llanwenog. But it took longer to create the Llanwenog. The intention was to breed a sheep that would complement the dairy herds which provided the main income for most of these Cardiganshire farms. Mixed grazing is good for the pastures because sheep's closer grazing makes the grass tiller out and the sward thicken; sheep's hooves level the surface after cattle have plunged it with their much bigger hooves; and a sheep flock tidies up the grazing after the cattle. Also it reduces the burden of worms that each species carries because they can ingest one another's worm larvae without being infected as they are destroyed in their digestive systems.

They sought a moderate-sized, placid, prolific ewe with good wool that was disinclined to stray and would produce prime butchers' lambs, entirely from grass. It was crucial that it lambed early, before the dairy herd was turned out in spring so as not compete with the cows for spring grazing. They also sought a ewe that would breed its own replacements, so the flock would become acclimatised to the terrain, and reduce the risk of importing disease.

It was a tall order to try to reproduce all these attributes in one breed. But in their own unobtrusive way the breeders achieved their objectives. The Llanwenog remained largely confined to the Teifi Valley until 1963, when the breed took four out of the six prizes in the class for small pure-bred lowland flocks in the national lambing competition organised by *The British Farmer and Stockbreeder*. The winning flock averaged 213 per cent lambing, a thing almost unheard of. Then the breed won again in the next two years with another flock achieving 230 per cent lambing; and for four consecutive seasons, between 1966 and 1970, John Hughes's flock at Cwmere, shepherded by his daughter Christina, averaged 217 per cent lambing. No other breed or hybrid could average more than 200 per cent.

This astonishing fecundity is an unforeseen consequence of the way the flocks have been managed. To reduce competition with the dairy herd for grass during the summer, lambs that were ready for market were sold fat straight off their mothers. Inevitably singles and twins would be ready earlier and be the first to go. The triplets and quads would be smaller at birth, would grow more slowly and tend to be the animals left at the end of the season. It was from these that the flock replacements would be chosen. As the tendency for multiple births is inherited, the flock would come to comprise ewes with that propensity. Ewes with these qualities admirably suited the dairying regime of West Wales: multiple births from medium-sized ewes meant that they needed to keep fewer ewes over the winter which did not graze the fields bare. And by lambing early in the year their need for grazing coincided with the flush of spring grass when there was plenty for both the lambing flock and the dairy herd.

Another benefit is the Llanwenog's longevity, due to the

ewes' excellent teeth. The ewes regularly have six or seven crops of lambs and so breeders can be choosier about their replacements and can pick a few of the best from each year's crop of lambs. When each ewe has twins, only about twenty ewe lambs are needed each year (10 per cent of all lambs) to maintain a flock of 100 ewes. The remaining 180 or so are available for sale, either for meat, or, if they are good enough, to other farmers for breeding. Also, having a closed flock, the ewes are acclimatised to their farm and fewer die from disease. The only thing the breeder needs to buy is a couple of fresh rams each year to keep up the flock's vigour. This is how to make lowland sheep farming pay.

Flock No. 1 in the Llanwenog Flock Book (the first flock registered out of the eighty-eight original flocks that formed the Breed Society in 1957) belongs to Huw Evans from Alltgoch in Llanwenog parish. Huw's great-grandfather and his two sons (Huw's grandfather and great uncle) came to Alltgoch in 1905 and the Unitarian minister at the time lent them the money to buy their first sheep.

The place exudes old-world, unostentatious order, hallowed by long and loving use. A spring flows peacefully from a pipe beside the house into a drinking trough, which feeds a large duck pond at the bottom of the yard.

Alltgoch means red forest, or forest red, I suppose it should be. And *goch* is a Welsh word borrowed from the Latin *coccum*, meaning 'scarlet' (the same derivation as 'coch' in cochineal). The River Coquet (red river) has the same Welsh root. The Celtic word connects geology and terrain with their effect on the quality of the sheep. Alltgoch and a few surrounding farms in Llanwenog parish overlie limestone, which is markedly different rock from the predominating Silurian mudstone that makes up much of the coastal rocks of West Wales. Limestone has a high natural pH and here contains deep mineral intrusions with ores of lead, zinc, even gold, as well as many other trace elements. Alltgoch produces high-quality livestock (particularly Llanwenogs) for the same reason as the mineral-rich soil around the headwaters of the Coquet produces fine Cheviots. This oasis contrasts strongly with the soils overlying the more acidic and predominant Silurian rocks in this part of West Wales, which need regular liming to raise the pH high enough to unlock their fertility and release their trace minerals.

In the forty years between 1957, when the Society was formed, and 1997, the Evans flock at Alltgoch won the large flock competition twenty-two times. They took the trophy thirteen years in succession between 1985 and 1997 and have never bought a sheep apart from the Llanwenog rams they have used for over 100 years.

Huw Evans is a Welsh bard who has been awarded three bardic chairs – not just of the metaphorical academic variety, but real, light-oak chairs, with arms and beautifully carved high backs. They sit in his parlour, his holy of holies, along with the cups, photographs, flock books and a century of Llanwenog memorabilia. He has written Welsh poetry since his teens and recited it at the National Eisteddfod. For the Royal Welsh Show County Appeal for Ceredigion in 2010, he teamed up with the artist Aneurin (Aneurin Jones) to produce a montage of a painting above a poem in Welsh, in praise of Wales and the ancient roots of its people. It is a real expression of belonging to the land and a paean to the beauty of the soil and its fruits.

Here are preserved the treasures and memorabilia of the things that have nourished the roots of four generations of Huw's family in West Wales: the land, the Welsh language and its poetry and Llanwenog sheep. He has every flock book, going back to the formation of the Llanwenog Society, in which are recorded the pedigrees of every sheep ever registered with the Society, farm by farm, flock by flock, family by family. These records are like the deeds of a house: they verify an animal's descent.

Beside the door hangs a tapestry map of Alltgoch made to commemorate Huw's birth on 22 January 1958. The loving labour that has gone into this – each field is embroidered in a different pattern – and the prophecy in it, or perhaps the pre-destination, set a seal on the course of his life. It is as if he was born for the farm and the farm for him, his destiny recorded by the tapestry, but not determined by it. That was the course set for him before he was born and Huw felt he could no more change that than he could alter the colour of his hair or his ancestry. His freedom and his happiness lay in his accepting this inheritance.

✌ 12 ✌

THE DOGS

Come my auld towzy trusty friend;
Waur gaurs ye look sae douth and wae?
D'ye think my favour's at an end,
Because thy head is turning grey?
Although thy feet begin to fail,
Their best were spent in serving me;
An' can I grudge thy wee bit meal,
Some comfort in thy age to gi'e?
For many a day frae sun to sun,
We've toil'd and helpit ane anither;
An' mony a thousand mile thou'st run,
To keep my thraward flocks thegither.

James Hogg, 'Address to His Old Dog, Hector' (1807)

O NE STILL AUTUMN AFTERNOON WHEN THE CLEAR
air was blue-sharp and the bracken golden on its
way from green to russet, I was gathering up stray
sheep a couple of miles away from the farm. I spotted them
a long way off, across a stream, grazing in fields on the other
side of a small steep valley. They had escaped through a newly
appeared gap in the wall. As it was at least three miles round
by road and would have taken me half a day to fetch them,

I decided to try and get them back the way they had gone across the beck, over the wall and to the top of the steep field where I was standing.

From high up on my side of the valley I sent my two dogs very wide, down over the beck in the bottom and up into the fields on the other side. The sheep hadn't seen the dogs coming, although they had heard me whistling, sensed something was afoot and conveniently flocked together on a hillock in the middle of the field. There were about thirty of them and the dogs got behind them before the sheep knew what was happening. As the dogs were working a long way off and it was too far to shout, I was directing them from my vantage point by whistle. I had never trusted either dog to work so far away from me before. When I lost sight of the sheep and the dogs in the valley bottom I had to trust to the dogs' good sense and training and hope they would bring the sheep up the steep slope and over the brow.

Nothing happened for a long time. I couldn't hear anything – no barking or baaing – and in the still air the noise of my heart pumping when I held my breath obscured any sound that I might have picked up from below. As I stood still, allowing the breath to trickle in and out of my open mouth, alert for the slightest noise, I was startled by a shallow cough from somewhere behind me. I spun round to see my neighbour Hilty Hope's cap peering above the top of the drystone wall, quietly observing proceedings. (Country people, particularly shepherds, do a lot of observing.) Hilty had a reputation for being good with a dog and had enjoyed some success at the sheep dog trials. At this stage we didn't speak; I lifted my hand but Hilty didn't acknowledge it.

A few minutes later, although I would not have admitted it, I was more than a little surprised, and quietly delighted, to see a small troop of sheep breasting the bank below, coming straight up towards me, with a dog in control at either corner.

As they gently brought the sheep towards the gate, I whistled for the dogs to lie down and hold the steaming sheep in the corner while I went over to have a word with Hilty. Neither of us spoke until I got up close to him and then he volunteered, 'That's a couple of useful dogs you've got there!'

Coming from a Cumbrian, particularly Hilty, this was high praise. Cumbrians don't readily dish out compliments. But then he added, 'Aye! It's as they always say, it takes a lazy man to have a good dog.'

Not only have British breeds of sheep been exported all over the world, but so have our sheep dogs. The black and white collie dog has spread everywhere across the pastoral world from its beginnings in the Scottish Borders. James Hogg wrote in 1809:

> A single shepherd and his dog will accomplish more in gathering a stock of sheep from a Highland farm than twenty shepherds could without dogs, and it is a fact that without this docile animal the pastoral life would be a blank. Without the shepherd's dog the whole of the open mountainous land in Scotland would not be worth a sixpence. It would require more hands to manage a stock of sheep, gather them from the hills, force them into houses and folds, and drive them to markets, than the profits of the whole stock were capable of maintaining.

In the past, in many parts of Britain, shepherds' dogs often accompanied their masters to church. An Edinburgh minister was taking the service one Sunday in a remote country kirk where dogs formed a substantial part of the congregation. When the minister rose at the end of the service to pronounce the blessing, to his surprise, the congregation remained seated. He looked around waiting for them to rise, but none moved. At length the clerk looked up from his desk below and shouted out 'Say awa', sir, it's joost to cheat the dawgs!' Experience had told them that if the congregation stood, the dogs thought the service had concluded and would get up to leave, disturbing the solemnity of the occasion with various noises and stretching.

There was some difference between drovers' dogs and shepherds' dogs. Often longer-legged and with naturally short tails (called 'self-tailed dogs'), the former had a smoother coat, and if anything were more indefatigable and cleverer than the shepherd's collie. Forever on the move, they were usually of the barkable kind rather than the quieter collie dog and, crucially, were happier driving away from their master than gathering towards him. The drovers directed their dogs and drove on their sheep and cattle by whistling and their dogs were easily as sensitive to their master's whistle as the shepherd's collie.

For upwards of 300 years, until the railway finally finished the droving trade, drovers and their dogs were involved in an astonishing animal migration and feat of endurance, moving huge numbers of livestock from the remotest parts of the British Isles – the north of Scotland, Ireland and the west of Wales – to the fattening pastures of the Midlands and East Anglia or straight to Smithfield market. The drovers regularly carried large sums of money, often as gold coin, and some (particularly the Scottish ones) were armed with pistols or swords to ward off outlaws and armed gangs who would try to steal their cash or their livestock. These men were resilient and resourceful and many were strong characters, with dogs of similarly reliable and robust temperament. Sir Walter Scott's grandfather was a drover who amassed a considerable fortune from the Scotch cattle trade at the end of the seventeenth century and beginning of the eighteenth.

There are many stories of their dogs' intelligence and courage. Progress with a large flock of sheep or a herd of cattle was often painfully slow, for they would graze the roadsides or open commons as they passed along. And once their initial burst of speed had been exhausted, the animals had to be encouraged to keep moving and that is where the dogs came into their own, tirelessly pushing the animals along and keeping them together and out of side-roads. Another of the dogs' duties was to make a way through the slow-moving animals to allow vehicles to pass. The dogs were so accustomed to doing this that on many occasions they would do it automatically and not have to be told.

It was not unusual for drovers to return home by carriage and send their dogs off to find their own way back. There are numerous reports from travellers of drovers' dogs making their way back north along the route they had travelled south. They would call to be fed at the inns and farms where the drove had rested on the way down and the drover would pay for their food next time he was going south again. Most dogs, to a greater or lesser extent, have an instinctive understanding of landscape and memory for places, but drovers' dogs had this capacity to a remarkable degree, it having been bred into them over many centuries. Sheep were regularly walked from the Scottish Borders south for fattening and once a dog had made the trip, rather like a London cabby doing the Knowledge, it would remember where the lanes and cul-de-sacs were and all the places the flock ought to be kept out of, and hold them to the straight road all the way.

A Welsh drover's dog, named Carlo, travelled back from Kent to West Wales on his own after his master had sold the pony he had ridden on the outward journey and arranged to return by coach; he fastened the pony's harness to Carlo's back with a note requesting the innkeepers along the route home to feed and shelter his dog. Carlo was well looked after and completed the journey within a week. His refreshment would probably have consisted of the traditional sustenance for drovers' dogs, bread and beer.

It is hard to explain to anyone who has never worked a

sheep dog how clever these pastoral dogs are and how completely the livestock farmer relies on them. Before there were so many motor vehicles travelling at high speed on rural roads, it was common to find farmers walking their flocks and herds from field to field, and even to market. I remember my neighbour walking his draft Swaledale ewes from their fell heaf to Cockermouth auction for the annual sale in October. The journey was about eight miles, and he did it in two stages, stopping off overnight about three-quarters of the way there, where the sheep could graze and rest.

I could move a flock of sheep on my own anywhere with just two dogs. One would stay behind with me, chivvying the sheep on, and the other would, when given the signal, push through the hedge or over the wall and go round the flock to head them off and hold them up if necessary. I had one dog, Ben, who was particularly good at this; the only thing he couldn't do was open a gate, but he would hold up the flock beside a gate until I could get there to open it. He wasn't infallible, but he was as reliable as many humans would have been.

Sometimes he got it spectacularly wrong. One day I was moving about 300 sheep through the village and had sent Ben on ahead to cover the gateways and junctions and to keep the flock on the road home. As we passed each side-turning he would be standing or lying in the way, blocking the sheep's entry. Occasionally he would bark as they passed just to hurry them along and remind them who was boss. At the end of the village, just before the gate of the field I intended to turn the sheep into, the owner had left her drive gate open. Ben went ahead and lay down in the road ahead of the flock to hold them up until I got there to open the gate. But this caused the whole flock to escape into her large garden, where they milled and wheeled about trampling everything in their path. By the time I had got them out the devastation was dreadful. I knew the woman who owned the house and she was very decent about it, generously admitting that she had been wanting an excuse to remodel her garden and had never got round to it.

Collie dogs are astonishingly astute to pick up a word or a whistle and act on it. I had a bitch called Tess who could not hear the word 'smoky' without immediately leaping to her feet, rushing outside and barking to chase the cat we kept with that name. Even if 'smoky' was hidden in the middle of a sentence such as 'it's very smoky in here', or 'smoky bacon crisps', she would never fail to pick it up, even if she seemed to be asleep at the time.

There is a story from 1846 told by a traveller to Scotland about a crofter's dog employed to keep animals off his small patch of oats and potatoes. The dog was lying by the fire in the house, where his master and the visitor were talking, and in the middle of a sentence concerning something else, without altering his tone of voice or stressing the words, his master interposed, '… the cow is in the potatoes …' and carried on with the rest of the sentence. The dog, which appeared to be asleep, leapt through the open window, scrambled onto the low turf roof of the house and finding that the cow was not in the potato field, went to look in the byre, where she found the cow safely tied and came back into the house. After a while the crofter spoke the same words again and the dog repeated his action. But when the words were uttered a third time, the dog got to his feet and, wagging his tail, looked searchingly into his master's face, as if to ask, 'Why are you trying to fool me?' and went back to his bed by the fire.

The meaning of *collie* is obscure. In *Waverley*, referring to events in 1745, Sir Walter Scott describes packs of 'collies' maintained by each village whose duty was to harry the often exhausted post-horses to keep them moving on from one parish to the next. The prosaic and often-cited derivation of collie is that it comes from 'coaly' meaning black; but that ignores the obvious fact that very few collies *are* black. Most have white, or tan, in their colouring. There is even a ginger, or 'red', strain favoured by certain shepherds, but shunned by others because at a distance it looks like a fox. Certain writers have suggested that as the sheep the dog herded were black-faced, i.e. 'coalies', the dog took its name from its charges. But nowhere are Blackface sheep referred to as 'coalies' and collie dogs shepherded other types of sheep, particularly white-faced Cheviots, long before Blackface sheep came to Scotland.

The more compelling explanation is that like many words to do with the pastoral life that have come down to us in modern English, it is derived from the language of the Celts and not from a later Latin or Germanic root. Collie seems to be one of those words preserved from an orally transmit-ted culture that has always existed apart from the dominant written culture of the educated ruling class. *Coelio* in Brythonic (the language spoken by Celtic people who occupied the south of Scotland, north of England and Wales) means *to trust* or *be faithful to*, and seems apt to describe this most trustworthy and trusting of dogs. In the Isle of Man, into the nineteenth century, the word for a sheepdog (not a Border Collie type) was *coly*, which seems to come from the same root.

The modern type of Border Collie largely descends from Old Hemp, a dog bred in 1893, by Adam Telfer, in the Nor-thumbrian Borders hills. His father was Roy, a flamboyant dog with a 'free eye and frank expression', and his mother was a more sensitive, shy bitch, called Meg. By the time he died in 1901 Hemp had fathered over 200 puppies with Telfer's own bitches and those brought to him by other breeders. These were then inter-bred so that the modern type of Border Collie became fixed, and within forty years there was hardly a collie that did not have some of Old Hemp's blood in its veins.

There is a long tradition (probably as old as pastoralism itself) of naming sheepdogs with short names of no more than two syllables in order to give a note of authority to the voice and ensure a quick response when the dog is called. The Roman writer Columella (from Cadiz), who wrote *De re rustica*, suggested two-syllable names such as Lacon, Ferox, Celer, Lupa, Cerva and Tigris. But British shepherds prefer one-syllable names, such as Bess, Tess, Roy, Nan, Bob, Jim, Ben and so on. To compare the effectiveness of this, the next time you are out in a howling gale with a companion, try calling 'Nan' and then 'Ferox' and see which gets lost on the wind.

Queen Victoria was fond of collies. She had been given one by Tim Elliot's grandfather, and when it died in the 1850s she journeyed from Floors Castle, where she was staying with the Duke of Roxburghe, to Hindhope, where Elliot presented her with a replacement The route of her pilgrimage was marked by the planting at intervals along the way of seedlings of the newly discovered Wellingtonia. A fine specimen still stands in the garden at Hindhope.

As successful and well-known as the Border Collie has become over the last century or so, it is by no means the only type of pastoral dog. There are many hill shepherds who do not rate the fancy 'creeping' type that is good at sheepdog trials, preferring a more free-ranging dog that can work on its own initiative. In England the hairy Sussex was the dominant breed before it was bred for show and turned into the Old English. It was a bold, resilient dog that could bear the summer heat much better than the Border Collie, which is happier in cooler weather. The Welsh Collie works with its head and tail up and is a more self-reliant type that can move large numbers of sheep across difficult open ground. It will bark and drive, and there is also a strain that will work close to, with a few sheep. Many are 'blue merle', with blue eyes, or wall-eyed, with one blue and one brown eye, which makes a very attractive, somewhat enigmatic dog.

There were strong types of dog that were bred not to gather the whole flock, but to catch and hold an individual sheep. The shepherd would point out the sheep to be caught and the dog would chase it and grab it by the neck-wool and hold it by opposing every movement the sheep made to break free. Sometimes the dog was strong enough to turn the sheep over and pin it down with its paws until the shepherd arrived. In the Isle of Man these were called 'houl'ers' because they would 'houl' on' to the sheep. The dogs of the crofters on North Ronaldsay do this on the shore to catch and draw out single sheep for slaughter.

In the pastoral countries of Europe and Asia there are many, probably older, types of dog that are used for guarding *and* herding. The Calabrian or Pyrenean Mountain Dog, or Illyrian, is a big woolly-looking creature that could easily see off wolves. Spanish shepherds, working for the Mesta, had big fierce dogs to herd and protect their huge Merino flocks from wolves during the bi-annual transhumance along the *cañadas* between Extremadura and Andalusia in the south and Leon and Castile in the north. Similarly protective dogs are found on the Steppes: the Hungarian Komondor (whose white coat of ringlets at a distance resembles a fleece and acts as camouflage when it is amongst the flock); the Puli is a black version of the Komondor.

In New Zealand they have created a version of the drover's dog called a Huntaway. Huntaways do not gather sheep *towards* the shepherd as a Border Collie instinctively does by eyeing them, but they drive them by barking and rushing about from side to side. They are powerful, tireless dogs, with short hair coloured rather like a Rottweiler. This type can drive large flocks of sheep across rough country, where an eye-dog like a Border Collie would be useless and are just the kind of dog that mountain shepherds use in difficult terrain.

The Kelpie is its Australian counterpart, bred to deal with large flocks and herds in extreme climates and difficult terrain. It is supposed to have some dingo blood from way back, but nobody would ever admit to crossing their dogs with dingos, not least because it was illegal in Australia even to keep a dingo as a pet because of their untameable instinct for savaging sheep. The Kelpie has tireless energy and complete devotion to its master. Its speciality is running along the backs of sheep to break up a large flock to get it moving. In this it is similar to the old type of Sussex sheepdog that could move a flock of sheep by 'mounting up' – running along their backs and snapping at their ears. The Kelpie's primary instinct is to keep sheep and cattle from moving away from its master, which it can do from a long way off, but it can also drive a large flock tirelessly over a great distance. Without these dogs the pastoral lands of the New World would, as James Hogg said of Scotland, not be worth a sixpence.

Unlike the rest of Europe, there have been no wolves (or other serious predators of sheep) in Britain since the Middle Ages and so shepherding is less a guarding and more a gathering exercise. British flocks on open land are encouraged to range by themselves over their whole grazing area and for this the shepherd needs fast dogs with stamina. There are two main types of sheepdog now found in Britain: the silent creeping close-to-the-ground Border Collie type that 'shows a lot of eye' – fixes the flock with a stare and dominates it by force of will; and the more active type of dog with long legs, that will get its tail up, bark when necessary and run about a lot, like the old drovers' dogs. Although both types can work a long way off, controlled by shout or whistle, the second is more likely to think for itself and act on its own than the Border Collie, which requires almost constant direction, even though it is a more precise instrument of its master's will than the rough-and-ready type, which wouldn't be much good at sheepdog trials, but would give short shrift to any insubordinate sheep.

In my experience – I've had both types – the second type tends towards extremism and wants to gather the whole flock, or none of it. Both types are fundamentally different from the guarding type of dog, which does little gathering because the flock tends to follow the shepherd with the dog making up the rear. These guard-dogs are more of a friend to the sheep, whereas the Border Collie is like a policeman, bred to master the wilder kinds of sheep found on the hills and mountains of Britain.

While I was working for Harry I paid £100 for my own pedigree collie pup from a renowned breeder in Westmorland – I named her 'Tess' – and trained her myself, or rather she taught me to how to train her. She was a natural. She had a half black and half white face and was wall-eyed, with one blue eye and one brown eye. This is an inherited trait in certain collies and considered by some to be an indication of superior ancestry. Wall-eye is from the Old Norse *vagleygr* meaning a film over the eye, like a cataract.

She was very fast and quiet over the ground and completely reliable at holding sheep in a corner, she ran instinctively wide round the flock, in whichever direction she was sent, and infallibly rounded them up and brought them to me. I quickly discovered that once I had trained her to stop when I told her to, I was three-quarters there. If I gave the signal of one long whistle, she would immediately clap to the ground. The next whistle was a short sharp note that got her to move on again. The knack to training her was to use the appropriate whistle when she was already doing, or about to do, what I wanted. She was quick on the uptake and soon began to associate that whistle with being told to do a particular action. Consistency and calmness were the watchwords. If she did something wrong I would whistle her back to me, and then get her to do it again. I avoided raising my voice unless it was to make her hear me a long way off.

Harry let me try her out on some of his geld (barren) sheep in the in-bye fields and occasionally sent me off with her to gather up a few stragglers. But because she wouldn't bark – and was 'a creeping wall-eyed bitch', as Harry affectionately referred to her – she was of limited use on the open fell, where she proved feeble at moving large numbers of sheep over hard country, particularly through waist-high bracken. They simply weren't frightened enough of her to move when she crept towards them, or they couldn't see her and because she wouldn't bark they didn't know she was there. The best she could do with Herdwicks on the fell was annoy them. Sometimes a ewe with lambs would call her bluff and run at her. She then resorted to biting, which Harry frowned on, because it showed she had lost her authority.

None of her keen pedigree impressed Harry because she wouldn't bark and wouldn't 'get up' – that is, she worked close to the ground with her tail down with the tip slightly curled up, not gaily on long legs with her tail waving like a flag. He said she was no more than a 'trial dog', a prima donna that would let you down on the open fell just when you needed her the most. He was right about the trial bit, and that she was not much good at moving large numbers of sheep, but he was wrong about her letting you down. She was the only dog I ever owned that never gave in. You hear about dogs watching over their dead master's body for impossible lengths of time – well, I'm certain Tess would have been one of those, loyal to the end.

Harry's judgement was partially vindicated one hot June morning. At that time of year when the ewes are in full fleece with active lambs at foot and the days long and warm, we had to set off very early to gather before the heat of the day. The commoners on our part of the fell had established a practice of leaving at four in the morning, just as it was getting light. We were each supposed to gather an area of the common and take the sheep to the communal sheepfolds below Force Crag, at the foot of Coledale Hause, at the head of the valley, where they would be sorted into their separate owners' flocks. Once they were sorted, their owners would leave at intervals with their sheep to avoid them getting mixed up with their neighbours'.

Harry and I were to gather The Side, on the north side of Coledale, which is a long, sweeping shoulder of land about 1,000 feet from top to bottom and two and a half miles long. In summer the steep slope is covered in chest-high bracken and the dogs find it very hard to 'lift' sheep out of the deep vegetation. Harry's friend Jack was with us, a lean, tireless, wily, cantankerous old Lakeland shepherd, who was well over seventy, with equally tireless, long-legged, barking dogs – black, white and tan versions of the Huntaway – which were doing a wonderful job (as he kept telling us) of ferreting out the sheep from the thickest bracken.

Some of the ewes were crafty enough to lie with their lamb in a bracken bed, flat to the ground, staying still until we had passed. Then, when they judged it safe, they would scurry away downhill, behind us, hidden in the bracken, until they were out of range of the dogs, and then emerge onto open ground half a mile away, running like crazy with the lamb sticking to them like glue. If they lay still they were very hard to find in the bracken, and an eye-dog like Tess could make little of them. I was soon reduced to shouting, whistling, clapping my hands, beating the bracken with my stick, and even barking myself, as Tess wouldn't, to get the sheep moving.

It was after seven by the time we met up with the others at the sheepfolds at Force Crag. Early shafts of sunlight penetrated the purple shadow under the crags and pierced the steam rising from the woolly cacophony. We had gathered up a huge flock of six or seven hundred ewes and lambs. They were hot and tired, and most of the ewes had become separated from their lambs and they were all making a tremendous noise, ewes baaing for bewildered lambs and lambs crying an octave higher for their mothers. We had them in a huge pen, roughly fenced with posts and wire netting and none too secure. It was the joint property of all the commoners and therefore nobody took responsibility for repairing it, everybody blamed somebody else for its decrepit state.

We let them calm down a bit and had a smoke before we began to put them through the shedder. A shedder is a marvellously simple and effective device for sorting sheep. It exploits two of the sheep's instincts: to follow the one immediately in front, and to move uphill if possible. The shedder is combined with a race, which is an alleyway with blind sides (ours were made of second-hand corrugated zinc roofing sheets), just wide enough for one sheep to pass along it at a time and sloping uphill if the ground will allow. So long as the sheep have a clear view up it to the other end they will think they can escape by moving along it towards the exit.

At intervals, set in the sides of the race and attached to posts, are little solid wicket sorting gates which can be swung from side to side to direct the sheep into separate pens. Our race had three such gates we could operate to divide the flock into four with only one passage along it. Some have two sorting gates hung opposite to one another in such a way that one man can work them both at the same time. This is pretty advanced sorting and hard to do without practice, because once the sheep gain momentum they just keep running and have to be directed somewhere. If the sorter loses concentration, rather like juggling, everything is lost and the operation has to start all over again.

Much of the language relating to sheep is either Celtic or, in certain places, Norse and has been little affected over the centuries by Latin or Norman French. *Gimmer* is a Norse word, as is *rake*. And one thing that leads directly back to the Celts is the method of counting sheep, supposed to have been used in Cumberland and Westmorland. I have never met any

Lake District (or other) shepherd who counted like this. Even 100 years ago, when Canon H. D. Rawnsley wrote *By Fell and Dale* (1911), the best he could do was find old men who could say that their fathers told them that their grandfathers always counted that way. The system is vigesimal (based on twenties) and remarkably similar to counting in traditional Welsh (and Breton) and apparently certain North American Indian tribes. The following is the Cumbrian version:

1 *Yan*
2 *Tyan*
3 *Tethera*
4 *Methera*
5 *Pimp*
6 *Sethera*
7 *Lethera*
8 *Hovera*
9 *Dovera*
10 *Dick*
11 *Yan-a-dick*
12 *Tyan-a-dick*
13 *Tethera-dick*
14 *Methera-dick*
15 *Bumfit*
16 *Yan-a-bumfit*
17 *Tyan-a-bumfit*
18 *Tethera-bumfit*
19 *Methera-bumfit*
20 *Jigget*

I think I learned from my mother that this was how the old shepherds used to count and every time he reached jigget the shepherd would transfer a pebble from one pocket to another. Some carried a tally-stick which they notched with their knife.

In traditional Welsh twenty-one is 'one-on-twenty' and thirty is 'ten-on-twenty', forty is 'two-twenty', and so on up to half-a-hundred, and then in twenties up to a hundred. Up to twenty it is so similar to Cumbrian counting that it seems likely to have stemmed from the same root, but the bigger numbers were forgotten as such counting fell out of use. I once asked Harry whether he had ever counted like that and he replied dismissively, 'Never! It's the kind of thing "townies" come out with.'

As Harry and I were going the same way, after we had sorted them we amalgamated our flocks and set off home together. The track from the sheepfold to the top of Coledale Hause was wide, very steep and rocky, and the sun was well up by the time we left. The stronger sheep quickly paired up with their lambs and set off at some pace up the hill. Harry put his dogs round them regularly to check their progress and keep the flock together, but they still got well ahead of the rest of the flock. The slowest were the smaller lambs, which were not only distressed because they had lost their mothers, but exhausted by the melee, the heat and the climb.

One after the other they would flop down onto their flanks, panting and refuse to move on. We started to collect up as many as we could carry at a time and took them in relays a hundred yards or so, to the top of the steepest part of the path, and laid them on the turf, then went back for more. Ewes kept doubling back, running from lamb to lamb trying to sniff out their own. Lambs were screaming for their mothers, dogs were barking and we were shouting, whistling and waving our sticks. And thus the little caravan moved slowly up the steep bare fellside towards the saddle of Coledale Hause. The younger, stronger sheep and those without family obligations had steadily drawn away from the main flock, which stretched out in a straggled line.

As the ground levelled out onto the hause I caught sight of the leaders making strongly across the flat land towards the top of Grasmoor a mile or more away and those ewes that had found their lambs were scurrying along behind them.

'We'll lose them if you don't get past them. Let that young bitch away!' yelled Harry.

I'd never seen him so discomposed. The sweat ran freely down his face and behind his ears and from time to time he took off his cap and swabbed his face and matted hair with it. Down the back of his light blue kytle (rustic cotton jacket) a broad slick of sweat had darkened the material to indigo.

Tess had been keeping close to me for the last half hour or so. She had given up trotting and the best she could manage was a doleful plod. She had worn herself out with her earlier efforts and her long, narrow, pink tongue lolled extravagantly from one side of her mouth, dripping moisture. She was also a little awed by the indefatigable long-legged fell dogs which barked regularly, rushing here and there, waving their tails like demented tour guides. Poor Tess was bred for elegance, and she seemed affronted by their coarse manners.

I shouted, 'Get away!' and waved my arm in the direction I wanted her to go round the flock. She set off well, but after about a hundred yards she slowed down and I whistled my 'Get-a-bloody-move-on!' whistle, which usually got her going, like flooring the accelerator in an automatic car. She disappeared into some 'slack ground' and I waited for her to come back into view on the other side of the hollow, but she didn't reappear. I whistled a few more times, but there was no sign of her. I was no more willing than she had been to run in the increasing heat, but when Harry shouted, 'Go and see what's up with that daft bitch!' I stumbled to the place where she'd disappeared from view and 300 yards away I spotted two ears pricked up above a grassy mound, and facing away from me. She had not heard me because I was running into the wind, so I managed to get very close before gasping for breath and screaming, 'Tess, you lazy bitch, you've let those sheep go!'

She flattened herself like a kipper, low to the ground, and as I neared she rolled onto her back with her legs bent, tail between her legs and started to whimper. I would like to say that I took pity on her because she must have been really exhausted, but to my shame I whacked her twice on her ribs with my stick. She yelped to her feet after the second blow and ran a short distance away with her back arched and tail now so far between her legs that she could have held its tip between her teeth. By this time the vanguard of the flock had reached the last climb before the summit of Grasmoor at least a mile away.

It would have been foolish for Harry to send a dog past them and risk chasing them over the top, even if his dogs still had the energy to do it. The crags on the west side of Grasmoor are treacherous and dogs and sheep could have fallen 1,000 feet and been dashed to pieces. Harry decided to send his dogs round part of the flock and leave the others gradually to graze their way back onto the heafs we had gathered them off earlier that day. We had no option but to come back and try to gather them another day, piecemeal, and on our own, because the next communal gather was not until the autumn. Not even if he enlisted my help, and Jack's, would we be capable of gathering the whole of these fells on our own.

Harry made it quite clear that he thought me to blame for this debacle. He deliberately ignored me when I caught up with him with the few sheep I had managed to keep hold of. I trudged on, downcast, feeling ashamed. We crossed the wide saddle of Coledale Hause, tussocky with old anthills and coarse sedges, and went on over long beds of loose, flat stones where the Skiddaw Slate had breached the surface and been broken up into barren scree over innumerable winters by rain and frost. We funnelled the flock across the path that winds between Sand Hill and Eel Crag and down into Gascale Gill.

Then he turned towards the flock, whistled to make his

dogs bark and pushed them on, shoosh shoosh, shooshing, and tapping the rocks with the metal ferrule on the end of his stick. As we picked our way down the path amongst the boulders in Gascale Gill I looked back over my left shoulder and there, high above us, I could see the escaped sheep slowly winding their way back across Coledale Hause towards the pyramidal peak of Grisedale Pike. We snaked down the loose stony track which followed the edge of the Liza Beck to Lanthwaite Green. Ice-cold water broiled over blue boulders and swirled into dark foamy, bubbling pools. Jack and his tireless dogs were silhouetted against the sky, high above us in the crags where he had been scouting for stragglers. I could just make out his slight figure hundreds of feet above and hear him whistling his dogs on. Harry shouted and waved, gesturing for him to keep going and join us at the far end, and his shouts echoed round the ravine. Jack stood motionless, peering down, cupping his ear, trying to catch what Harry was saying, like an ibex on a rock ledge, silhouetted against the limitless blue afternoon summer sky.

Gascale Gill broadened out and descended steeply towards Lanthwaite Green where it joins the valley and where the path turned right behind the fell wall towards the farm. Our sheep were now too exhausted to do more than plod on disconsolately. We had to take it slowly to avoid any of them collapsing in the heat. All the farms adjoining the common had wicket gates in the fell wall through which their owners could gain access onto the fell. Whenever we were near one of these gates we would slip through any sheep that had flopped down and refused to go any further, to wait until we could get back to collect them with the van. It was after two when we finally reached the intake behind the farm with a flock of very weary sheep and exhausted dogs. My legs and knees ached and I would have found it hard to walk much further, but Harry seemed remarkably fresh for a man in his mid-fifties who had climbed up and down 2,500 feet in the heat of June, covered at least fifteen miles and been on the go for more than ten hours with nothing to sustain him but spring water and a packet of Embassy Regal.

After we had put the flock safely through the fell gate, Harry stopped in the path and waited for me to draw level with him; it was clear he wanted to say something. He considered me sideways from under the cap he always wore at an angle, and declared with a sigh, 'I told you that creeping bitch would let you down.'

The dogs were remarkable for their stamina. Fell shepherds rarely allowed their dogs meat. I remember hearing some theory that giving them meat would turn them into sheep 'worriers' because they'd get a taste for it. They were usually fed on *euveka* (a dialect word for cooked flaked maize) and cow's milk. I don't know what nutritional value it has, but I was surprised it was enough to sustain a dog doing the work these were expected to do. But that was the way their fathers had fed their dogs and it was good enough for them. I'm the first to see virtue in tradition, but not uncritically following a bad one. Many fell shepherds treated (and for all I know still do treat) their dogs quite harshly. They depended on them so fully that it was as if they had to treat them cruelly so as to even things up in some way. Perhaps I was being sentimental, but I always felt sorry for their dogs which gave their all and got so little in return.

After I'd been farming for about a year I saw an advert for Bearded Collie pups for sale in the *Scottish Farmer*. I phoned the seller, who lived in Lanarkshire. I told him I wasn't keen to travel all that way on a wild goose chase; he assured me that they were working dogs and even offered to send me a bitch and if I wasn't happy I could just send it back. He wanted £75 including carriage. Two days later a carrier dropped off a cardboard box about the size of a case of wine. It was hard to recognise the pathetic scrap inside as an eight-week-old

Bearded Collie pup. It was about the size of a Yorkshire terrier. We got the terrified creature out of the box and set her on the kitchen floor next to a bucket of that day's eggs. She immediately took possession of them, bared her teeth, growled and snapped to keep us away and then gently picked out an egg, without breaking it, and cradled it between her paws, by turns licking it and uttering menacing growls if she thought any of us was going to approach. She was clearly starving but too frightened that we would take it from her to allow herself to eat the egg. We left her alone for a few minutes and when I peeped round the door she was wolfing down the last of three eggs, shell and all, which she almost immediately sicked up.

Our vet said she had the worst infestation of intestinal worms he had ever seen – and lice – and mange. It was the only time I have ever considered reporting anyone to the RSPCA; instead I rang the man and told him he had cheated me and I wanted my money back because it had cost more in vet's bills to put the dog right than it was worth. He was defiant. He didn't accept there was anything wrong with the dog and he said I had got what I paid for. I told him he ought to be ashamed of himself, neglecting a dog like this. 'Well send the bitch back!' he said and put the phone down.

Eventually we pulled her round, but she never fully recovered. Apart from being about half the size she should have been, she always had small, half-grown weak teeth which weren't improved by her obsession with picking up pebbles in the yard and dropping them at visitors' feet for them to throw for her. Her early neglect had induced such paranoia that she would not eat with other dogs or people, or even cats, nearby. She was so terrified that her food was going to be stolen from her that she would stand over it uttering bloodcurdling snarls and baring her teeth. By the time the other dogs had finished Nan had eaten nothing, so busy was she defending her dinner. Even when she was fed well away from rivals, the slightest hint of competition caused her to gulp it down, growling and chewing at the same time.

She took me by surprise by coming into season at an absurdly young age and being impregnated, or 'lined' as Harry indecorously described it, by a dog belonging to a neighbouring farmer – I can't remember whose. She gave birth to a litter of five puppies and was a marvellous mother. Her offspring astonished us. They grew to be big hairy clever Bearded Collies, twice the size of their smooth-coated mother. It would have been so easy to have had her put down, because she was a useless dog, and untrainable because she was so psychologically disturbed. But she bred a litter of the best dogs I ever had. She never had any more pups and never grew any bigger than a medium-sized spaniel.

I kept a dog from the litter and we called him Spot. I sold the rest to neighbouring farmers. Spot grew to be a powerful hairy black, tan and white dog. I could rely on him to gather a whole hillside of sheep while I walked along the bottom, encouraging him by shouting and whistling. But every dog has its faults. Perhaps because Spot was the cleverest dog I ever owned, he was also the most temperamental. If I shouted at him too harshly he would immediately stop what he was doing and go straight home. He was just too touchy to be a reliable working dog. I could almost hear him saying *I'm not being spoken to like that; you can gather your own bloody sheep if you can't keep a civil tongue in your head.*

He was the best dog I ever had at holding up a flock of sheep in the road. Nothing ever got past him. He could even do the same thing with cows if I told him in a particular tone of voice to 'Watch them!' With bullocks, or cows with calves, which can be difficult to stop once they've got their tails up and start to run, he developed the technique of going straight for the leading animal's nose and biting it hard. It seldom failed to stop them in their tracks.

Another of his tricks, if you wanted to hurry on a herd of cattle, was to leap up and nip the tail of the last cow. He knew how to avoid getting kicked by being off the ground before he bit. Sometimes he would hang on near the top of the cow's tail and brace himself with his front paws against its rump. Once he'd got them moving he would run back to me delighted with himself, wagging his tail and panting with pleasure.

But he was an all-or-nothing kind of dog which could only gather every sheep in a field or none at all. He would set off close to the boundary, aiming to go right round the perimeter, and often be out of sight for many minutes, working on his own initiative. I knew he would either come back with the entire flock or, if he didn't reappear, he had come across some thing he couldn't deal with on his own. But if I raised my voice he would stop in his tracks, lift his head, look at me just to be certain and, having satisfied himself that his ears were not deceiving him, set off for home at a determined trot. Occasionally – and it was only occasionally – I could coax him into staying and carrying on with his work. But only if I made an extra-special effort to persuade him I hadn't meant what I'd just said. With the right soothing tones he would sometimes grudgingly accept my craven retraction. Sometimes it meant starting again from scratch because the sheep had taken the opportunity to get away while Spot and I resolved our differences. But even after I had apologised, he was usually still on his dignity. I had to speak quietly and show him the utmost solicitude, or he would respond to the slightest further raising of the voice by trotting off home without even pausing to check whether or not his ears had deceived him.

Once he had decided to leave it didn't matter how far from home we were (I had some land about three miles away, across the River Cocker), he would usually make it back before me and I would find him lying in the yard as if nothing had happened. 'That dog's the boss of you,' said Harry after he'd witnessed one of his walking-off-the-job episodes. I suppose he was, but I loved him for his charm and his intelligence and, I suppose, because he was impossible.

One morning he seemed particularly recalcitrant. I coaxed him out of his bed and he dragged himself across the yard into the field behind the farm. Usually when he saw some sheep he would work. But this morning I had never seen him so unwilling to run. I couldn't think what I'd done to upset him. I set him off round the sheep to fetch them into the yard. He moved off slowly and seemed to be ambling round them, taking his time. I thought he was being lazy, so I whistled my 'get a bloody move on' whistle. This seemed to encourage him to move a bit faster up the slope, but after less than a hundred yards he flopped down, panting, on the grass. By the time I reached him he was lying on his side, apparently exhausted, and refused to get up. He licked my hand as I lifted him up and then I noticed that his gums and tongue were yellow. As I was carrying him back to the house I saw the whites of his eyes were yellow too.

'Leptospirosis' was our vet's telephone diagnosis, which he confirmed when he saw the dog twenty minutes later. 'Weil's disease! They catch it from rats' urine; either a bite or an open wound. It destroys the liver. Like hepatitis. There's nothing much I can do.'

'Will he survive?'

'I'm afraid not. If I'd caught it early on I might have been able to save him, but it's too far gone.'

Then I remembered that the dogs had been chasing rats in the buildings a few days earlier – which they often did for fun. I heard Spot yelp and come running out trying to shake off a squealing rat which was hanging by its teeth from his upper lip. I thought nothing of it at the time because I'd never heard of Weil's disease.

The vet gave him an injection. Spot had always hated injections and tried to bite anyone who went near him with

a hypodermic needle, but apart from a whimper when the
needle went into the scruff of his neck, he made no protest
and I knew then he must have been very ill. I took him home.
He never moved from his basket by the Aga, apart from the
odd flick of his tail when I spoke his name. He died quietly
after lunch the next day.

I spent the afternoon digging a deep grave in a quiet place
at the bottom of the garden. Carefully cutting the sod into
squares and setting them aside, I dug off the topsoil and put
it into one heap, before digging into the sub-soil, taking care
to keep the sides straight and neat. It was a proper grave, fit
for the best dog any man could have wished for. Afterwards,
when I was able to think a bit straighter, it seemed appropriate that he should have died at that time because it turned out
to be the last summer we lived at the farm and Spot couldn't
have settled anywhere else. He belonged to the place, no less
than the Herdwicks to the fells and the sessile oaks and carpets
of bluebells in the spring belong to the thin gravelly soils. He
came to me from that land and it was right that I should have
had to give him back to it. It is proper that his bones should
lie there.

After I'd buried him I climbed to the top of the high land
behind the farm called Brackenthwaite Hows. It was early
April, the time of year when the land starts to dry out after the
mud of winter; when the hawthorns green before the pungent
blossom comes cream with magenta anthers. My mind was
numb, overwhelmed with the poignancy of spring and grief
for the loss of my dog. Brackenthwaite Hows is a lump of
slightly harder rock than the surrounding Skiddaw Slate. It has
resisted being scoured out by the glacier that formed the valley
during the last ice, and sits about 250 feet above the surrounding land at the intersection of the Buttermere, Lorton and
Loweswater valleys. On the top the bare grey rock is scored
by deep striations caused by the scraping of boulders trapped

in the glacial ice, as it moved inexorably down the valley to the sea.

To the west the ground falls steeply away through Scale Hill Woods to Crummock Water and the River Cocker. To the east the slope gently declines to Lanthwaite Green, on the gravel delta that has been washed down by the Liza Beck from the surrounding fells over centuries. From the top is the best panorama for the least effort anywhere in the northern Lake District. Up Loweswater valley to the west, past Crummock Water to the high central fells in the south, and into the Vale of Lorton and on to the Solway Firth to the north. To the east the looming slopes of Whiteside and Grasmoor rise 2,000 feet steeply from the valley floor. This is the country I come from. And if I belong anywhere on this earth, it is here. I am comforted to know that the best dog I ever had lies in this soil.

✨ 13 ✨

THE SACRIFICIAL LAMB

The sacrifice of God is a troubled spirit: a broken and contrite heart, O God, shalt thou not despise.

Psalms, 59.16

THE DETAIL OF THE FOOT-AND-MOUTH EPIDEMIC OF 2001 is now hazy in my memory. But I do retain a series of vivid images that come to me whenever I think about that sinister time when almost half the livestock of northern England and southern Scotland was slaughtered and cremated on huge open funeral pyres. Three of these pyres were built within five miles of where I live. One was only a mile and a half away to the west, across the fields, and I can still see and smell the smoke drifting for days on end. There was hardly a place in the English countryside not overblown by smoke from the pyres like warning beacons, whose flames lit up the dark night. The smell of roasting flesh, burning diesel and creosote blew everywhere, even into our cities.

I remember the stream of lorries laden with dead cattle coming past the house, and the charred cloven hooves pointing to the sky from the flames. 'Bio-security' checkpoints were set up at strategic places on remote country roads where you had to drive over mats impregnated with disinfectant or where, even in the middle of the night, men in boilersuits would emerge from a Portakabin and spray the underside of your car with pungent disinfectant before you could pass on.

People accepted all this with goodwill and co-operated because we were told it was a national emergency and they wanted to do their best to help to bring the killing to an end. But the public officials soon lost control of the slaughter programme and had to bring in the army to get them out of the mess. Soldiers did their duty on their usual pay, while all the private contractors involved were making a fortune. The slaughtermen had never seen so much money. One farmer said they were 'having the time of their lives. They seemed to get a kick out of killing on a grand scale.' Hauliers had never been so busy. There weren't enough slaughtermen or vets in the country and they had to import them from abroad. Many had never seen foot-and-mouth before and erred on the side of caution, diagnosing the disease in animals, particularly sheep, that only had a skin infection. One farmer told me that his sheep were slaughtered before they even knew the result of the test, but by the time it came back negative it was too late to save them.

In fact much of the livestock did not have the disease, but were slaughtered anyway, in a 'contiguous cull', the idea being to create a buffer zone between the disease and the supposedly healthy stock. If a farm was chosen for culling, every cloven-hoofed thing on it was slaughtered, down to the children's pet lamb or goat. The killing was ruthless and shocking and lasted for weeks. Most farmers were well compensated; some never went back into stock farming – took the money and retired, or changed direction. Quite a few bought property, or built new barns; more than one bought a holiday villa abroad. Not a few wives took the once-in-a-lifetime opportunity to divorce while there was money in the bank.

At the time I found it hard to understand why some farmers did everything they could to resist the slaughter and were devastated when they were forced to surrender their stock. I could understand that pedigree breeders, and those with heafed hill flocks, whose animals could not be replaced, would find money a poor substitute, but for commercial farmers getting all that money at once didn't seem a proper cause for dejection.

Now I think I understand it better. It was partly the waste that shocked and upset them; they were sickened by the profligacy of the destruction. They reared their animals to feed people, not to have them incinerated on huge pyres. Many had bred and nurtured their livestock over generations, they were their whole life, and when they were destroyed in a day they suffered a wrenching bereavement.

The slaughter came at a dark time for farming, when prices were so depressed that farmers were selling their sheep, wool and cattle at less than the cost of production. There is no doubt that the compensation saved many struggling livestock farmers and transformed the businesses of many more, injecting a good deal of capital into the countryside. But there was also a barely expressed, because it is hard to express, but deeply felt, unease about the whole business. Coming, as it did, so early in the new century, there was feeling that the epidemic and the slaughter held a sacrificial, even apocalyptic, significance and, in some vaguely understood sense, was a blood sacrifice. But for what?

I do not know whether there was any deeper element of sacrifice about the slaughter, but I do know that many of the animals lost their lives unnecessarily. A pedigree sheep breeder told me that none of his sheep had the disease until he found a pair of slaughterman's white hooded overalls that had been thrown over the hedge into one of his fields. Then less than twelve hours later he had a phone call from DEFRA to inform him they understood his farm might be infected and they were coming to test. Others, miles from any infection, found cow tongues that had been thrown into their fields amongst grazing animals, and they received the same quick response from DEFRA that resulted in their animals being slaughtered.

We might think that sacrifice, particularly blood sacrifice, belonged to a primitive, pre-rational world that has been abolished by modern enlightenment. At the heart of sacrifice is the offering to a divinity, in propitiation or praise, of something of great value to the offerer. Although there are bloodless sacrifices, of artefacts, food or drink (libations), the most powerful and enduring have always involved the spilling of blood, preferably innocent blood, and very often, in the Abrahamic tradition, the blood of sheep, preferably lambs.

The mythical and mystical references to sacrificial sheep in our history and culture go back into prehistory. The lamb, especially a white lamb, is the symbol of renewal, tenderness, sacrifice and innocence. In Jewish, Christian and Muslim symbolism the lamb is the perfect sacrificial victim. The Jewish sacrifice of the Passover lamb, the Islamic sacrifice of a lamb at Ramadan, and the sacrifice of Christ, the Good Shepherd, who died on the Cross as the ultimate sacrifice to redeem the world. Christianity fulfilled the Old Testament prophecy that a suffering messiah would be like a lamb led to the slaughter.

But blood sacrifice as part of a religious rite has been practised by nearly all cultures across the world from very earliest times. This impulse for blood sacrifice would appear to be something deep within humanity which requires expression and which if denied in one form will find another. And the one perennial victim of blood sacrifice in Europe and Asia has been the poor sheep.

Although the God of the Old Testament required animal sacrifice, it was only as a temporary covering for sin until he should send his Son to shed his blood as the ultimate, perfect sacrifice to end all sacrifice, for as St Paul said, 'without the shedding of blood, there is no forgiveness' (Hebrews 9:22). Animal sacrifices were commanded by God so that the individual could experience forgiveness of sin. The animal served as a substitute – that is, the animal died in place of the sinner, who was thereby forgiven, but only temporarily; that is why the sacrifice needed to be offered over and over again. But for Christians animal sacrifice ceased with the coming of Christ as the ultimate sacrificial substitute for all eternity (Hebrews 7:27).

At the heart of her great account of travelling in the Balkans in 1937, *Black Lamb and Grey Falcon*, Rebecca West describes the sacrifice of two black lambs on a rock in the Sheep's Field on the Plain of Kosovo, the sacred place of the Serbs. The ancient rock, where sacrifice was made, had candles stuck in every crevice and was stained red-brown with the blood of a thousand years of sacrifice and gleamed with the drying blood of the cockerels and lambs that had been killed there during the previous night.

A father circled the rock three times, then kissed it, and handed up his little daughter to a man standing on the rock, climbed up himself and watched as another man took a little black lamb to the edge of the rock and drew a knife across its throat. A jet of blood spurted out 'and fell red and shining on the browner blood that had been shed before'. The father had caught some of the blood on his fingers and with it he made a circle on the child's forehead. The sacrifice was then repeated.

A bearded Moslem standing by the rock explained that the father was doing it because his wife had got his daughter by coming to the rock and sacrificing a lamb and 'all children that are got from the rock must be brought back and marked with the sign of the rock'. West was revolted by what she saw as an abominable practice 'of shedding innocent blood to secure innocent advantages'. Of inflicting pain on 'something weaker than themselves'.

She shuddered at the primitive belief that 'if one squares death by offering him a sacrifice, one will be allowed some share in life for which one has hungered. Thus those who had a lech for violence could gratify it and at the same time gain authority over those who loved peace and life.' She was disgusted that the slaughterer of the lamb was very well pleased with the importance his cruel act had given him and she thanked God that Christianity denounced this cruelty, 'the prime human fault which the military mind of Mohammed had not even identified'.

England has long prided itself on its concern for animals. Even though an animal was to be slaughtered, English law required that it be treated as humanely as possible and spared any unnecessary distress and suffering. This usually meant animals were stunned before slaughter, either electrically, or with a captive bolt fired from a pistol that entered the brain and retracted, killing them before their throats were cut. But for many years an exemption was allowed for ritual religious slaughter, which was intended to satisfy Jewish demand for kosher, or *shechita* meat. This exemption allowed the killing, by cutting their throats without stunning, of about 1 per cent of the sheep then slaughtered in Britain. Many were uneasy about cutting an animal's throat without stunning it, but as long as it was a minor exception to the rules, it was tolerated.

The Jewish tradition was adopted, almost without modification, by Islam. In Arabic an animal sacrifice is called *dhabihah* or qurban, and in Islamic law is the prescribed method of ritual slaughter of all animals. It consists of a swift deep incision with a sharp knife across the throat which is intended to cut the jugular veins and carotid arteries, but leave the spinal cord intact. The animal has to be fully sensate when its throat is cut and while it is being bled to death, otherwise its meat is *haram* (forbidden) as carrion.

When the first Muslims came into Britain they took advantage of the kosher exemption to provide themselves with halal meat. However, over the years, as the Muslim population of England and Wales has increased (now 5 per cent) so has halal slaughter, so that half of all the sheep now killed in Britain are slaughtered in accordance with the prescriptions of Islam.

Abattoir owners say that about 90 per cent of these are stunned first by passing an electrical current through their heads, which renders them temporarily insensate, but does not kill them. But this is a suspiciously round figure, and Rizvan Khalid, a director of Euro Quality Lambs, a Muslim abattoir, was honest enough to admit that even though he says most of the lambs killed in his abattoir are stunned, he would prefer non-stunning. This does not give much confidence that stunning will always be done if the inspectors are not looking. For most Muslims pre-stunning satisfies the Islamic requirement that the animal should be alive when it is bled. Most of the other 50 per cent of the sheep that are killed in accordance with the animal welfare regulations are also back-stunned, which involves passing a second electrical current through their body which stops the heart, or shot with a captive bolt. Muslims will not accept this because they consider that it kills the animal and makes it *haram*.

On 1 January 2014 the EU Regulation 1099/2009 *Welfare of Animals at the Time of Killing* ('WATOK') superseded the old 1995 *Welfare of Animals (Slaughter or Killing) Regulations* ('WASK').

Article 3 of WATOK requires that 'animals shall be spared any avoidable pain, distress or suffering during their killing and related operation'.

Article 4 (1) says 'Animals shall only be killed after stunning in accordance with the methods and specific requirements … set out in Annex 1.' The methods of killing referred to in Annex

1, which do not result in instantaneous death (simple stunning) have to be followed as quickly as possible by a procedure ensuring death, such as bleeding etc.

However, where animals are slaughtered by 'methods prescribed by religious rites' the requirements of paragraph 1 do not apply provided that the slaughter takes place in a slaughterhouse.

The regulations are in force across the EU, but national governments are free to make their own rules that add to the protection of animals and do not reduce the effect of the regulations. Poland, Norway, Iceland, Switzerland, Sweden and New Zealand have banned slaughter without stunning and Finland, Denmark and Austria require the animal to be stunned immediately (almost simultaneously) after its throat has been cut if it has not been stunned before.

In Britain, the Farm Animal Welfare Council, which exists to advise ministers how to promote humane treatment of animals, recommended that slaughter without stunning should be banned, as did many other bodies, including the British Veterinary Association and the European Federation of Veterinarians. They all condemned as cruel the killing of an animal by cutting its throat, because it caused 'avoidable pain, distress and suffering'. The process of bleeding out takes time: the animal has to held immobile for at least twenty seconds for sheep, and two minutes for cattle, while it bleeds to death, and the regulations dictate that it is not supposed to be moved during this period. This need to give each animal a period of individual attention makes it difficult to automate the process.

Research by Compassion in World Farming also highlighted a phenomenon slaughtermen call 'ballooning'. After the throat is cut, large clots of blood can quickly form at the ends of the carotid arteries, slowing the rate of blood loss and delaying the drop in blood pressure that causes the merciful loss of consciousness. As many as 62.5 per cent of calves suffered from ballooning. John Webster, professor of animal husbandry at the University of Bristol, emphasised that in religious slaughter fear afflicts an animal just as much as pain. 'What is totally unacceptable is the distressing fact, for the cow, that she is conscious of choking to death in her own blood.' Nonetheless, the exemption for religious slaughter remains.

For those sheep that are only head-stunned there is no practical difference between slaughter that is halal or haram. But there is a spiritual difference which is important to Muslims. In order to be halal sheep have to be killed by either a Muslim or 'a person of the Book' (a Christian or a Jew) and, whoever does it, he must dedicate the animal's life to Allah by reciting over it as it is dying, 'In the name of Allah, Allah is great.' Some insist that during the process the animal is also turned east, to face Mecca. Others insist that the animal must not be killed within sight of another animal, or see the knife that is about to cut its throat.

It is not as if the adherents of the two religions which seek to take advantage of the exemption can agree amongst themselves on how to kill animals. Orthodox Jews insist on *shechita* meat, Reform Jews do not and the Board of Deputies of British Jews is quoted by Nick Cohen in an article in 2004 for the *New Statesman*, saying that the overall attitude of Judaism is best summed up in the twelfth chapter of the Book of Proverbs: 'A righteous man regardeth the life of his beast.'

Islam is even more confused. A recent report on halal slaughter by EBLEX, the body that promotes the beef and sheep industries in Britain, highlighted the widely different understanding of halal slaughter by everyone involved. Of the forty-one abattoirs doing halal slaughter, only fourteen were willing to participate in the survey, raising the suspicion that there were more abattoirs that did not stun than would admit to the practice. There was no generally accepted definition of halal slaughter. It certainly did not seem to turn on whether the animals were stunned, because many abattoir owners stunned their animals because they admitted that it was better for their welfare. It is certainly not prohibited by Islam to stun them, so long as they are not considered to be dead when they are bled. Many consumers were prepared to accept the word of a Muslim butcher that the meat was halal. It seems that the essential factor was that the slaughterman be a Muslim and say the correct prayer over the dying animal.

There is no Koranic prohibition on stunning. For example, all lamb slaughtered in New Zealand has to be stunned first, yet it sells huge quantities of sheep meat to the Muslim world. In Britain the Muslim market for halal sheep meat is about 20 per cent of the whole but the relatively small share of the market that is non-stunned meat seems to be increasing. Many people I spoke to were unwilling to talk about it, even though they thought it was cruel. Some farmers were anxious not to spoil the Islamic market for their old ewes, which they feared would collapse if they caused too much fuss about not stunning. But as stunning is acceptable to all but a small hard-line element in Judaism and Islam, its prohibition could hardly be interpreted as oppressive to minorities, or racist. Rather, it would be perfectly consonant with European concern for the welfare of animals, some of which appears to be shared by most of those who insist on halal slaughter.

It is hard to find out whether or not the meat we buy *is* halal slaughtered, and even more difficult to know whether it comes from an animal that has been stunned before having its throat cut. Many in the meat trade, including many Muslims, feel that it ought to be made mandatory to label meat as halal, stunned or non-stunned, so that consumers can make an informed choice. After all the fuss about the alleged cruelty to the small number of foxes that were killed by dogs during fox hunting, it is disappointing, to say the least, that few are prepared to raise the same outraged opposition to the tens of millions of farm animals that have their throats cut without stunning in ritual slaughter every year. A disagreeable smell of double standards comes from the animal rights movement, which is against the shooting of pheasants, most of which are dead before they hit the ground, and their relative silence over religious slaughter. It is beyond argument that cutting an animal's throat without stunning causes 'pain, distress or suffering' that could be avoided if it was stunned first.

Non-Islamic consumers have little idea where their meat comes from and many British people seem not to care too much. In any case, as they can't quite reconcile the meat on their plate with the death of an animal, they would rather put the subject out of their minds. The disturbing thing about it is that many people involved simply will not talk about halal slaughter; they seem inhibited to say aloud what they think, for fear of upsetting religious sensibilities, or being accused of being 'racist'.

The government and other authorities that could ban slaughter without stunning are no better at discussing it. Many Christians and, interestingly, Sikhs are firmly against throat cutting before stunning for sincerely held religious reasons. And many secular liberals passionately oppose it because they feel it to be needlessly cruel. The National Secular Society has been particularly vocal in its opposition to the suffering of the animals involved. In the end, it comes down to the question of which sensibilities, religious or otherwise, are to take precedence, because they can't all be given equal weight. In the meanwhile, much suffering that could be avoided goes on.

❧ 14 ❧

THE MODERN AND
THE FUTURE

The more alfalfa [Major Major's] father did not grow the more money the government gave him, and he spent every penny he didn't earn on new land to increase the amount of alfalfa he did not produce.

Joseph Heller, *Catch 22*

FARMING FACES FORMIDABLE CHALLENGES. THOUSANDS of farmers, as small primary producers, are economically weak in the face of a few powerful buyers and traders. Farmers have little control over the prices they receive and to survive they have to use considerable ingenuity in keeping down their costs and finding new ways to use their land. This is nothing that they have not faced and conquered in the past, and they are astute at finding ways of dealing with it.

Over the last half-century the national flock has become younger and more prolific as breeders have responded to the growing demand for lamb meat. They would once have kept geld ewes for their wool, and wethers for their wool and mutton, but these have largely disappeared from our pastures. This demand for ever younger, leaner and smaller lamb carcases seems a profligate waste of sheep's lives, requiring ewes to have the largest possible litters of lambs, which are then killed at what our forebears would have considered absurdly young ages. Many never reach six months, and few make it to a year old. Consumers are missing out on the delights of eating grass-fed mutton, which is a superior meat to the often bland lamb that supermarkets offer, especially the kind of lamb that has been fattened on cereals.

But farmers can only produce what people will buy. So as tastes change breeders have responded by producing ever leaner and more muscled types; pushing the boundaries into the development of breeds that would have impressed and amazed Bakewell. This skill of pedigree breeders, and their readiness to innovate, makes British sheep still the best in the world and is nowhere seen to better effect than in the new meat breeds of which the Texel and the Beltex are supreme examples.

Texels originally come from the island of that name, off the coast of Holland. The breed may have a Roman origin, and, like the other Dutch breeds that conquered the world in the second half of the twentieth century, the Friesian and Holstein dairy cow and the Landrace pig, they are practical manifestations of domestic livestock that make no concessions to beauty. I am willing to concede that there is beauty in their functionality, although I'm not sure I can appreciate it, but they display such a depressing worldliness of purpose that the joy I could get from rearing a cussed old Herdwick, or a stately Lincoln Longwool, would elude me if I found myself shepherding a flock of Texels.

These are square barrels of meat with a leg at each corner, short neck, black nose, white legs and face bare of wool, and are very similar to the type Bakewell was aiming to breed, judging from pictures of his best rams. In fact, over the years, the old Dutch sheep that became the Texel is believed to have had an infusion of New Leicester at some time in its past. It is also said to have had the benefit of some Lincoln Longwool blood, but it is hard to see what effect it had because its tight medium-wool fleece is neither pirled nor lustred – nor long. Nonetheless they are probably the most popular modern meat sire for producing lean butchers' lambs, although they have not managed to eclipse the Suffolk, and it is doubtful they ever will.

As if the Texel weren't ugly and meaty enough, Beltex sheep were created in Belgium in the 1970s to provide the kind of lean meat the modern market expects, and almost every other attribute has given way to that. As its name suggests, Beltex is a portmanteau of Belgian and Texel and just about the most extreme manifestation of a meat-sheep that we now have. It takes Bakewell's vision of the ideal sheep to surreal heights of meatiness that transcend all connection with its terrain. It has been bred to satisfy the urban consumer's demand for meat, without compromise. It proclaims, 'You wanted meat, well here it is! Beat that!' There is a stark beauty in its brutalism. It is an honest sheep whose existence exposes the hypocrisy of a modern urban world that pretends it would like its meat to come from some lovable old breed.

It is like the residents of a charming market town resisting a Tesco supermarket that they say will ruin the independent shops, and yet as soon as it opens abandoning their wonderful butcher and fishmonger. It comes down to the perfidious pragmatism of the British that whenever they have to choose between standing on a principle that might be for the welfare of the people and making money, they invariably choose the money and convince themselves they are doing it for the best of reasons.

By the end of the last century the transition that began in the eighteenth century was almost complete. The value of wool declined to the point that it cost more to shear most sheep than their wool was worth. Progressive breeders, who could sense the way things were going, began to develop breeds, and types within breeds, that not only grew quickly to maturity but also needed less shepherding – particularly at lambing time. They recognised that the cost of keeping sheep was the crucial factor in determining profit because there was little they could do to increase the end price, either of meat or of wool.

One of these progressive breeders was Iolo Owen from Anglesey, who created a breed in the 1960s called, rather prosaically, the Easycare. He started with a type of Welsh Mountain (the Nelson), which he crossed with the Wiltshire Horn, our only native breed that sheds its scanty fleece in spring. He chose from within the Wiltshire breed rams that were naturally polled to ensure the resulting cross had no horns and did not need shearing. The ewes were easy to lamb and attentive and devoted mothers, and their lambs grew quickly from grass alone, without extra feeding. Strictly speaking, Evans created a type rather than a breed proper, because Easycare sheep are chosen purely on their performance rather than fixed breed characteristics. There is a breed society, but it discourages competitive showing because it fears it would lead to an emphasis on irrelevant points of appearance, such as shape of ears or colour of skin, and lose sight of what they believe to be the all-important intrinsic attributes that assessment by eye alone would never detect. The Easycare Sheep Society claims that the costs of shepherding are reduced by up to 80 per cent if you keep Easycare sheep. They have not caught on.

Another response to the declining value of wool and the increasing cost of labour is Andrew Elliot's Chevease. On his farm in the Scottish Borders he has crossed his Cheviot ewes with Easycare rams. The purpose of the exercise was to get his flock to shed their coats (they can hardly be called fleeces) in early summer and avoid having to shear them. He claims that the new type – arguably not yet a breed – is less likely to die from getting on its back (and suffocating from bloat), and does not need dagging, treating for fly-strike or castrating, because the ram lambs mature so early that they are fully fleshed before they are sexually mature. When the sheep are moulting in early summer, the fields are strewn with locks of wool, like the aftermath of a rock festival, and the sheep look very shabby during the shedding process.

Since the very low point in the price of wool that impelled Andrew Elliot to breed it out, we have seen the beginnings of a world shortage caused by a reduction in global flocks, and thanks to that and the skill of the Campaign for Wool, begun by the Prince of Wales and run by John Thorley, the indefatigable former Chief Executive of the National Sheep Association, and the writer Nicholas Coleridge, the price of British wool has more than tripled in the last few years, admittedly from a ruinously low price. But wool is such a unique and valuable substance that it is hard to imagine the demand diminishing further. It therefore remains to be seen whether abandoning wool was entirely wise. It was certainly courageous, but if it ever became profitable again it would be hard to breed the wool back. On the other hand, wool will have to increase a good deal in value before Andrew's decision will be vitiated.

These new types tend to be kept in pure-bred flocks with all their replacement ewes coming from within the flock. This brings the advantages of resistance to disease and acclimatisation to the terrain, but loses the benefit of hybrid vigour, which some breeders believe does not compensate for the advantages that come with the flock being acclimatised to its ground, such as resistance to disease.

Those breeders who were attracted to the Wiltshire Horn for its wool-shedding property, and its remarkable ability to survive all year on grass alone, got the added bonus that it has a strong natural tolerance to internal parasites, particularly worms. Tim White, who keeps 800 of them on the chalk plains around Monkton Deverill, in the depths of Wiltshire, has redis-covered this quality after trying all manner of eclectic breeds in his search for a low-maintenance grazing animal.

Over the years, many things have been tried to control parasites, including arsenic, mercury and nicotine, but it was not until pharmaceutical companies developed sophisticated wormers ('anthelmintics') that it became easy to kill them. But these medicines quickly became an expensive necessity, with sheep losing any natural resistance and needing to be 'drenched' (given a liquid dose) at least twice a year to control their 'worm burden'. Flockmasters did not know whether they were breeding from resistant animals or those that only survived because of the medicine. But Tim has taken a more radical approach. He does not aim to eliminate worms entirely, but to breed animals that have a high natural resistance to infes-tation. He tests their dung for the presence of worm eggs and culls those with a high count. Resistant animals will have about ten worm eggs per gram of faeces, whereas heavily infested animals will pass between 1,500 and 2,000 eggs per gram. By selecting from those that have the capacity to resist infestation he is creating a flock of natural survivors that cost as little as possible to keep and give the best return.

Being a grazier, Tim neither owns land nor is a tenant of any land for a term longer than a season's grazing. He simply rents grazing wherever he can get it and is proud that, apart from his flock, his only equipment is a battered pickup and trailer, a set of mobile sheep handling pens, his dogs and his hand-held computer, the 'black box' that contains all the ovine records which he uses to select his breeding stock. The great value of the Wiltshire, in Tim's eyes, is that they can look after themselves. He only goes round his lambing ewes twice a day – 'the less you bother them the better' – and he has virtually no lambing complications because there are no over-sized, overfed cross-bred lambs to give trouble at birth. This is one resourceful man's way of making a living from sheep farming. He might be criticised for harking back to a breed many thought outmoded, but I have a sneaking suspicion that in his ingenuity in finding land to farm and his low-cost system he is pointing firmly to the future.

There is another way of facing the future, but it points in the opposite direction. It's about 150 years since sheep were last milked in Britain – to make cheese. The last commercial milking flock is said to have been a Hill Radnor flock near Sennybridge, and before that Cheviots had been milked in the Borders well into the nineteenth century. But the guaranteed high prices paid for liquid cows' milk by the Milk Marketing Board after the First World War put paid to milking anything other than cows.

So when Crispin Tweddle set up his sheep-milking enterprise, fourteen years ago, at Orchid Meadow Farm, just south of Shaftesbury, there was not much competition. He and his wife only intended to buy a field to preserve the view from their country retreat. But they found themselves buying the whole of 180 largely derelict acres. He was not happy to accept the depressed returns from the dairying that was traditional on these damp valley farms with small steep fields, mild winters and early springs. Milk quotas, the ruinously low wholesale price and the throttling grip of the supermarkets meant there was little profit in milking cows.

Whenever dairy farmers struggled to survive because they were not getting a proper price for their produce, DEFRA's usual advice was to 'diversify'. But, for a farmer, diversification was like advising David Beckham to try rugby if his football wasn't going too well. Most diversification amounts to little more than persuading the farmer's wife to take paying guests. But milking *sheep* is a different matter. That *is* diversification.

There was only one drawback. The grinding servitude of twice-daily milking, and the dairyman's dependence on a monthly cheque, makes him not as free as the corn grower who, at least in theory, has enough capital to see him through a whole year. The cow keeper is like a wage-slave whose spirit is stunted because he can't lift his head long enough to see beyond the next milking. So, just as the corn grower despises the cow keeper, who looks down on the pig farmer, who patronises the battery egg producer, they all look down on the man who milks sheep. Milking *sheep*! How servile is that!

But the beauty of being a newcomer to farming is that he is either unaware of this subtle pecking order, or, like Crispin Tweddle, is rich enough not to give a damn about it. Also the newcomer has not had his enthusiasm dampened by knowing that one of the easiest, and often the pleasantest ways to lose a great deal of money is to put it into 200 acres of English farmland. It can be just as if you had opened up a shaft in the earth and were trying to fill it with money, there is simply no end to how much it will swallow. But the Tweddles were too shrewd to fall into this trap. Their timing was impeccable, getting in at the bottom of the market, both for land and for livestock.

They started milking 200 Dorsets because they were the local sheep and would breed at almost any season of the year. Having them lamb in three batches ensured a steady supply of milk throughout the year. Then they crossed the Dorsets with East Friesland rams and immediately increased the output. The flock is now nearly all Friesland, yet another of those breeds of livestock from Holland (or nearby) that have revolutionised British farming. They produce a lot of milk if they are fed large amounts of grain and the best hay or silage, but they most emphatically do not live off the land. They would breed at nearly any time of the year and their milk has a greater proportion of fat to protein, which is good for making yoghurt. Their delicate femininity belies a sturdy constitution. Interestingly, they owe their remarkable powers of milking and fecundity to an early cross between the East Friesland breed and Bakewell's New Leicester.

Orchid Meadow now has about 1,000 ewes in milk at any one time, kept in five flocks that lamb in sequence throughout the year to ensure a regular supply of milk, which is sold either as liquid milk or to make yoghurt, through Woodlands Dairy in Blandford Forum, which the Tweddles also own. Some also goes for making cheese. Demand far exceeds supply because there are only about 15,000 ewes being milked in Britain and producers can almost set their own price, currently about a pound a litre, roughly in line with the French price.

The huge number of lambs (about 230 from each 100 ewes) are taken from their mothers at about three days old and reared on powdered milk – inside in winter and outside in summer, by a Polish milkmaid. I felt sorry for the hundreds of these forlorn little pink creatures, sucking rubber teats attached to buckets, with not a ewe in sight. I know their fate is the same as lambs that suckle their mothers, but the process seemed so brutally industrial, similar to battery hens in cages, reducing the creature to its function stripped of all sentiment and consideration of beauty. I struggled not to find the system depressing, then I told myself I was being sentimental. But on reflection I do not think my unease arises from sentimentality; rather I felt it a shame that sheep, of all our domestic animals probably the least suited to industrial farming, should be reduced to it in this way.

I am not suggesting for a moment that this even approaches the industrial farming you see in feedlots across North America, or the millions of chickens reared in vast sheds, but this is not the fluffy-bunny business that the name Orchid Meadow and the word organic might imply. This enterprise contrasts vividly with Tim White's free-ranging Wiltshires because keeping large numbers of sheep in close proximity encourages parasites and disease. Foot rot and parasitic worms are the biggest problem. The bacterium that causes foot rot is more contagious in the warm (40–70°F) conditions in buildings and to control it the whole flock has to be walked religiously through a foot bath with a corrugated bottom (to open the cleats). But even more of a problem is that being organic precludes the routine use of drugs that would prevent the usual parasitic infestations, for example by the highly prolific Barber's Pole worm, *Haemonchus contortus*, which can cause heavy losses where sheep are kept intensively, particularly in young sheep, if the weather is warm and damp after a drought,

Although organic sheep milking is an innovative response to the difficulty farmers face trying to make a profit, it remains to be seen whether it will stand the test of time.

But the second threat to the future of farming is new and much more insidious. It is something not easily vanquished because it springs from a pervasive ideology that has stealthily taken hold of Western society over the last century, and which I loosely call environmentalism. While it affects all types of farming, its greatest effect is on sheep farming in the hills and uplands, and its proponents intend to have even more. When an idea like this strengthens into a belief that morphs into common currency, it cannot easily be displaced.

The average age of farmers is now about sixty. Many older farmers readily admit that they were only able to get into farming because of the opportunities that arose after the Second World War and it would not be remotely possible nowadays: there are hardly any farms to rent, farmland is priced out of the reach of nearly everybody except those who already have it or are very rich, and EU regulations and subsidies have ossified rural land use and closed off opportunities that would once have been available. One Cumbrian farmer, whose son has become a solicitor, put it with characteristic bluntness: 'They've taken the enterprise out of it. I can't even plough one of my own fields without asking for a "derogation" from DEFRA.' Many upland sheep farmers are now welfare claimants, prepared to jump however high the state tells them to, seduced by the easy money to be made from not farming.

Older tenant farmers cling on because they can't afford to buy a house to retire into and be left with enough capital to live on. The stewardship schemes have made this worse because farmers can draw an income for not farming – especially in the uplands. Many farmers are receiving more in a year to care-take their own farms than they could ever make from farming, without the work and the risk. This income pays the rent (if they are tenants) and the bills and they often live in a beautiful place in semi-retirement. Why would they want to leave when all they could afford would be what one called a 'street house' in the nearest town?

Throughout the centuries keeping sheep has been one of the ways into farming for young people with enterprise. They could start small and fairly quickly grow big enough to make a living, renting land wherever they could and taking any opportunities to build up a flock. Nowadays it is much more difficult because so little land is changing hands. It is not so much that the land is held in too few hands, rather that the interference of the state in the working of a proper market in farmland and tenancies is having a stultifying effect. As the average age of farmers increases beyond normal retiring age, it will become ever more pressing to loosen up the transfer of productive land to a younger generation for the national benefit. To allow a body of ageing farmers and absentee landowners to keep a grip on farmland and prevent young blood coming in is likely to be as ruinous for national food production and the social fabric as the view of a recent chairman of the NFU, Peter Kendall, that the solution to the impending shortage of milk is to create ever larger industrial farms. Where are the shepherds and farmers and countrymen of the future to come from? And who will look after our farmland?

But there is an even more extreme threat to sheep farming. I came to Pikenaze Farm on the Woodhead Pass to see a White-face Woodland flock and found instead another reason why sheep are leaving the hills and came face to face with the challenge that they have to meet. Whiteface Woodlands, named after the Woodland Vale in Derbyshire, are part of that great tribe of shortwoolled, white-faced horned sheep that traditionally occupied the west side of England into Wales in a great sweep from the Scottish Borders to the West Country.

Until the Rare Breeds Survival Trust became involved the breed was kept going in the high Peak by a few breeders who valued its virtues on its native moors, even when it was reduced to 1,000 or so breeding ewes. It is still a rare sheep, but not endangered in the way it was. It has strong bone and a whiteness to the face and legs and a pink nose that echo its origins in the limestone country, and its stronghold is the moors around the Woodhead Reservoir near Glossop, where Yorkshire, Lancashire and Derbyshire meet. This is a bleak place, of notoriously heavy snowfalls, blown by wild winter winds, and is frequently cut off in a bad season.

The moor is in an Environmentally Sensitive Area (ESA) and the farmer, the remnant of whose flock grazes part of it, is being paid not to keep sheep on a large area of it. He has sold 800 ewes (about half the flock) that formerly grazed a large area which has now been fenced off to be left to nature. He has agreed to do nothing with it – no grazing or anything else – for the ten years the scheme lasts. The brown dead fronds of *Molinia* grass ('flying bent') are thigh deep and the wind whips it off the hill and blows it into the reservoir, where it rots.

The fenced-off and abandoned moorland stood out from the rest, like a brown patch stitched onto the green background. Natural England says it is being 're-wilded'. But nobody could tell me why land that has produced meat and wool for many centuries is being left to revert to a state it has not been in for thousands of years. There have been sheep and cattle grazing these moors, probably since the Iron Age, and it is hard to understand the benefit of paying a farmer to let this land lie idle.

During the very bad snowstorms a couple of years ago, some of the remaining sheep walked on a snowdrift over the fence into the set-aside area. The snow was so deep that it was difficult to get them out. But hardly had half a day passed before DEFRA phoned the farmer to tell him that his sheep were 'in the ESA land' and he had better get them out 'double quick'. When he asked how they knew, when the road had been blocked with snow for days, the caller replied that their satellite had been watching the site and had seen the sheep.

As part of the 're-wilding' programme Natural England sprayed some of the moorland with weed-killer, churned up the surface and sowed it with heather seed. And ... nothing grew but weeds. The assumption is that sheep grazing moorland is bad for the 'natural' flora and if they are kept off, the native flora will return. But it is hard to know what *is* natural, especially after the thousands of years of grazing which has created the moors.

Across the valley, the largest peat bog restoration project in Europe, Moors for the Future Partnership, was formed in 2003 with the intention of re-creating the blanket bog that it was believed had once clothed the high moors of the Peak District and the Pennines. Ecologists claimed that the peat layer had been damaged by industrial pollution, sheep grazing, walkers, farming and weather erosion. The scheme involved first erecting a twenty-mile fence round a large area of the moor called Bleaklow. The sheep were expelled and work began to 're-water-log' the peat and cover the bare ground (which was said to be six square kilometres in 2003) with hundreds of tons of heather brashings and 'geo-jute' (netting made from unbleached jute fibre that is supposed to stabilise the surface and rot away after a season). They began work to block up the grips (drainage channels) by carrying stones up from the valley by helicopter.

Millions of pounds have been spent on this project, including paying the farmers not to keep the thousands of sheep that used to graze the moors. The partnership use a number of arguments to justify the vast cost: peat captures carbon and reduces the rate of climate change, reduces water run-off and flooding downstream, and 're-creates a unique ecosystem' that supports the 'unique wildlife of the moors'. They believe that by re-introducing sphagnum moss and cotton grass the peat and the blanket bog will regenerate. Most of the work has been done by helicopter. Up to the end of 2012 they claimed to have reseeded about 1,500 acres of moorland using 8 billion grass and heather seeds.

Nowhere do they mention how much carbon is emitted by the helicopters and all the other machinery they are using, or the vehicles the army of advisers uses, or the rest of the energy they are putting into this project. They admit that getting native species to grow is an extremely difficult and expensive process. Bilberry seeds will not readily germinate unless they pass through the stomach of a bird and they haven't 'found a safe way of replicating this process yet'. Sphagnum moss is also difficult to germinate and they try to persuade it to grow by a method called 'hydro-seeding', which is a fancy way of describing spreading seed mixed with water from a helicopter, or quadbike or knapsack sprayer. Depending on species, the seed costs between £155 and £3,000 a kilogram. Then it costs between £45 and £75 a hectare to sow it, not to mention the fertiliser, the 'monitoring', the testing, the research and a whole host of ecologists' salaries, pensions, vehicles, fuel, and so on and so on.

The heather brash is cut only from moors in the Peak

District National Park to avoid bringing ticks into this tick-free area. It has to be cut by tractor and forage harvester in autumn and winter after it has seeded, loaded into bags and lifted by helicopter and then spread across the site either by helicopter or by hand. Between 2003 and the end of 2009 they had spread 18,500 × 125 kg bags of heather brash. They have also rolled out vast lengths of geo-jute up and down the gullies and secured it with metal pegs that apparently 'bio-degrade' (rust away) quicker than wooden pegs in the acid soil. The idea behind it is to stabilise the surface of the peat in the gullies.

It is far from clear that the peat *is* disappearing, or that this project will result in the peat growing back. And if the peat *is* eroding, it might be part of a long-term natural process that cannot be reversed. It is by no means certain that grazing did cause the erosion of the peat layer or loss of the vegetation. Grazing with sheep generally improves a sward by making it knit closer together and tiller out. Certainly sheep grazing, coupled with burning, improves heather by keeping it young. The ecologists behind the project do not know what has caused the reduction in growth of sphagnum mosses. The scheme is based on certain presumptions: that sheep grazing is bad; that global warming, or climate change as we now have to call it, is happening and is having an effect on the moors. But it is far from certain that these *are* to blame. All the money and effort is really being spent on a huge trial-and-error basis, so far with little to show for it. But neither the organisers nor the armies of volunteers are discouraged. They say things like 'It's got to be worth it', 'The signs are good' and 'When you think of the future of our water, our wildlife and our atmosphere it's got to be worth it.' The slightest sign of regeneration of any native species is taken as a sign that their faith is justified.

These moors are being turned into something like a huge municipal park, in theory open to everybody, but in fact highly regulated and restricted. Humanity is excluded from large areas for 'the benefit of the ecosystem'. The whole effort seems to be some sort of spiritual exercise in a secular age. And it is far from certain that it will increase the prospects for wildlife, because dominant predators, such as birds of prey and foxes, are left unchecked in the belief that nature will balance itself if left to its own devices.

The project doubtless appeals to those who have absorbed the belief that farming is a wicked practice, is destroying the planet and is morally inferior to environmentalism, which is a Good Thing. That view has infiltrated large parts of the Western psyche without our fully realising it. The only difference between what might be called a middle-of-the-road, broadly green view and the re-wilders is a matter of degree. Your ordinary conservationist is more of a preservationist, wanting a kind of shaggy garden, where sheep may safely graze, tended by quaint rustics. But re-wilders want Nature to be unchained, freed from human influence, so it can achieve its soaring potential. Out would go farmers, tourists, ramblers; in would come a brave new world of wolves, bears, beavers, otters, reptiles and all that goes to make up a 'thriving ecosystem' with the European straight-tusked elephant, which apparently browsed the flora of Western Europe 115,000 years ago, at the top. Extreme re-wilders will not rest until this rough beast of a pachyderm, its hour come round again, is resurrected. I can see that going down well in Keswick.

A confident civilisation, with a future, cultivates its land and values its husbandry because these are on the side of humanity. It honours the trust, building on the work of previous generations, before passing it on to the next a little better than it found it. But re-wilders don't behave as if they have a next generation. They seem to see themselves as the last of the line, with a sacred duty to return everything to Nature, in one grand final gesture.

Environmental schemes and re-wilding projects have

already gone a long way towards ridding our hills of sheep. There's hardly a flock left in the Highlands, and they're fast vacating the rest of the uplands because farmers are being paid not to keep them. Not keeping sheep is such a profitable business, that like Major Major's father in *Catch 22*, getting rich from not growing alfalfa, the more sheep farmers don't keep, the more money they make from not keeping them, and the more land they can set aside for not keeping sheep.

It is true that we have become disconnected from the reality of the natural world and no longer know where our food comes from; environmentalism may be an understandable reaction to industrial farming, but it's not the answer. Letting the land run wild does not produce anything of tangible value that can be eaten, worn, traded or otherwise put to use.

Some will try to claim the fruits of re-wilding do have a value, but, as Harry Hardisty used to say, you can't eat trees. As the world's population increases and the nations grow ever more restive, how are we going to feed ourselves? Creating an Arcadian idyll, from which human activity has been banished, is a utopian fantasy that would have us retreat from large parts of our cultivated land, surrendering to the wilderness hard-won terrain and destroying many centuries of human effort. It's a fantasy that spits in the face of the future.

Sheep have been our constant companions since the dawn of civilisation, partners in taming the wilderness, and are the most versatile of our domestic livestock. They have adapted themselves to every nuance of climate and terrain. Their grazing has enriched our soil and made the beauty of our fields and farms. There is hardly a sheep-keeping country in the world whose flocks have not benefited from an intermingling of genes from British sheep. Sheep are in our national blood, they are part of the life of the land that still pulses beneath the surface of modern Britain. They are our pastoral heritage and our future. Let us celebrate them.

ᘐ GLOSSARY ᘐ

Agistment: the taking in of animals to graze for a fee. From the Old French *agister*, to lodge – hence *gîte* – and *jacitare*, to rest or lie, hence *hic jacet*. The person who takes in the grazing animals has a lien over them – i.e. is entitled to keep them until their owner pays the agreed fee.

Anthelmintic: a substance that expels or kills intestinal parasites

Bradford Count: an old measure of the fineness of a fibre of wool (now superseded by the micron). It is based on the number of hanks, 560 yards long, of single-strand yarn that the wool sorter judged could be spun from a pound of **top**. A count of 56 meant that the pound of top would make 56 hanks, i.e. 17.8 miles of yarn; see **micron**

Britch: in the north of England, a sheep's back end – i.e. its breech

Chilver: a ewe lamb before its first shearing (used in the south of England); – see **gimmer**

Chine: the ridge along the shoulder and backbone

Cleat: the gap between the hooves containing a gland which excretes scented lubrication to reduce chafing between the hooves. When the scent is transferred to the ground it encourages flocking.

Cowie: polled ram in the north of England

Crimp: the natural folding and curling of each wool fibre

Dagging: clipping away wool round a sheep's rear end that is either soiled or might become soiled with urine or faeces, to discourage blowflies laying their eggs there; see **strike**

Dinmont: an obsolete word for a shearling in the north of England and southern Scotland

Drench: to dose with any kind of liquid medicine

Down breeds: breeds created as terminal sires for meat production. They all descend in one form or another from John Ellman's improved Southdown.

Draft ewe: older female sheep that has been drafted, i.e. drawn out, from a flock for sale. Usually refers to mountain and hill breeds that go for further breeding with Longwool breeds to produce hybrid breeding females.

Ewe: mature female sheep

F 1 generation: *first filial generation*, i.e. the first-cross between two different pure breeds

Fold: a place where sheep are gathered together

Fell: any unimproved grazing land – usually unfenced common, but not exclusively

Finished: when a grazing animal has completed the development of muscle and begun to lay down fat and is well-fleshed enough to form an edible carcase; see **slow feeding**

Fly-strike: see **strike**

Foggage: the regrowth of grass after a crop of hay has been taken

Geld sheep: a sheep that is neither in lamb nor suckling lambs – a barren sheep

Gigot: a hind leg of lamb or mutton – from the French, but of unknown origin

Gimmer: a female sheep synonymous with girl; as in gimmer lamb (girl lamb), gimmer hogg (female that has been weaned but not yet shorn)

Hank: loose skein of wool 560 yards long

Heaf, heath, heft: part of an open hill where a sheep feels it belongs and to where it returns instinctively. This proclivity to become attached to a piece of land is instinctive (probably in all animals) and is also passed from generation to generation. Hill and pure-bred sheep often have the strongest attachment to their heaf.

Heterozygote: (adjective heteroxygous) a zygote (fertilised egg) formed from gametes with certain differing pairs of alternative characters, one of which is dominant and the other recessive; cf. **homozygote**

Hirsel: the area of land upon which a flock lives and also the number of sheep that a shepherd can comfortably manage on a hill. From the Old Norse *hirtha*, to herd or tend.

Hock: the backwards pointing joint in the hind leg of a quadruped between the true knee and the fetlock.

Hogg: also hogget, a sheep that has been weaned but not sheared for the first time. It can refer to either sex; hence it is usually qualified by using tup (male), gimmer (female) or wether (castrated male).

Hoggust: a North Country word for a building where hoggs were housed during their first winter as an alternative to sending them away to better land; cf. **agistment**

Homozygote: (adjective homozygous) a zygote (fertilised egg) that has inherited particular genetic characteristics from both parents; cf. **heterozygote**

Kemp: short, coarse, brittle, hairy fibres in the fleece that are not wool

Kytle: a short, coarse linen jacket, popular among farmers and shepherds all over the north of England, especially Yorkshire and Cumbria (from the Norse)

Lambing percentage: the number of lambs reared expressed as a percentage of the number of ewes in the flock

Lanolin: waterproofing fatty oil secreted onto wool and used in ointments and creams

Lisk: groin. In a sheep it is the loose skin between the hind leg and the abdomen.

Luck money: a small cash gift made by the seller to the buyer of livestock, like a tip, which seals the bargain and invests it with good fortune

Lustre: the natural sheen of certain Longwools caused by the shape of the cup-like plates that make up each wool fibre and enhanced by the excretion of lanolin

Micron or **micrometre:** one thousandth of a millimetre, one millionth of a metre; used as a measure of the fineness of a fibre of wool

Moorit: chocolate-brown (light or dark) wool

Mule: first-generation hybrid. In sheep it is the offspring of a Bluefaced Leicester ram bred with a Swaledale ewe. From the Latin *mulus*.

Nott or **not:** Southern English word for a shearling

Polled: hornless

Race: an alleyway only wide enough for one sheep to pass down at a time; see **shedder**

Raddle: see **rudd**

Rake: to range over, or spread across the whole area of a piece of land, or to drive ahead of; from *raka*, Old Norse, to drive or to drift

Ram: entire male sheep, i.e. one that has not been castrated; see **tup**

Recessive: an inherited characteristic which is masked by an alternative dominant characteristic and which will only become apparent when it is allied to the same recessive characteristic from the other parent; see **heterozygous**

Rise: the growth of new wool under the old fleece

Rudd or **ruddle** or **raddle:** the coloured marking fluid or crayon applied to a ram's chest which will mark a ewe's rump whenever the ram has mounted her and show that she has been served – also the act of marking with the fluid

Scurs: vestigial horns often little more than scaly growths where the horns would have been

Shearling: a sheep of either sex that has been shorn once; thereafter two-shear, three-shear, four-shear, etc.; see **nott**, **twinter** and teg

Shedder: a wicket gate hung in a race in such a way that by swinging it from side to side a flock of sheep can be sorted into two or more groups by directing them individually into separate pens, obviating the need to handle them

Slow feeding: description of a grazing animal that takes a long time to mature: see **finished**

Smit: a Cumbrian word for the fluid used to mark wool to indicate ownership or something else the shepherd needs to remember about the sheep.

Spain: to wean a lamb from its mother; pronounced variously 'spean', 'spyan'

Staple: the natural length of wool fibres in the fleece

Steading: collective term for the farm buildings, usually with the farmhouse

Stell: enclosure (usually circular, sometimes cruciform) usually formed by stone walls, built on a hirsel to which a flock can retreat for shelter and food in bad weather. Rarely found outside Scotland.

Stint: the owner of a stint has the right to put as many animals – sheep, cattle, horses or pigs – on a common grazing as the stint represents. Stinted pastures are where the number of animals allowed to graze is limited (stinted) by agreement between the common graziers or by the owner of the soil.

Store sheep: animals that are not finished but are at an intermediate stage between being lambs and being ready for the butcher

Strike: in conjunction with fly- or blowfly; when greenbottle flies have laid their eggs on a sheep and it is either about to be eaten or is already being eaten by maggots

Sward: the grass covering a pasture

Teg: sheep in its second year; see **shearling**, **nott** and **twinter**

Terminal: accompanied by sire or cross is a sheep involved in the final part of the sheep pyramid, producing lambs for their meat

Top: wool that has been washed and combed to make all the fibres lie parallel ready for spinning into **hanks**

Tup or **tip:** see **ram**

Twinter: a Cumbrian word for a sheep that has seen two winters

Wall-eyed: having one eye of a different colour from the other, usually one blue and the other brown; from the Old Norse *vagleygr*, having a film over the eye

Wether: a castrated male sheep; usually in conjunction with 'hogg' or 'shearling' or 'two-shear', etc. to indicate its age

᭟ LIST OF ILLUSTRATIONS ᭟

2 © Diana Steriopulos; 3 by kind permission of the *Farmer's Guardian*; 4 © Adrian Legge; 9, 10 © Louise Fairburn; 13 Robert Bakewell (1725–95), painted by Boultbee (1745–1812) used with permission of Charnwood Museum; 14 Lithograph by C. Hullmandel after J. W. Giles (1842). All other photographs taken by the author.

While every effort has been made to contact copyright-holders of illustrations, the author and publishers would be grateful for information about any illustrations where they have been unable to trace them, and would be glad to make amendments in further editions.

❧ SELECT BIBLIOGRAPHY ❧

The work of three writers has been invaluable to me: Professor M. L. Ryder's magisterial *Sheep and Man* is a marvellous modern encyclo-paedia of a lifetime's ovine knowledge; William Youatt's *Sheep*, from 1840, is unbeatable for its insight, range and perspective; and I have relied heavily on Robert Trow-Smith's highly readable two-volume history of British livestock husbandry. Trow-Smith's prose is a delight to read, possessing the enviable gift of making a compelling narrative from a potentially dull subject. Apparently his editor dreaded receiv-ing a manuscript from him because there was so little he could do to improve it.

The following is a selection readers might find helpful for further study:

Roy Baker, Alan Bull and Peter Lambley (eds.), *A Natural History of the Catfield Estate*, Norfolk and Norwich Naturalists' Society, Norwich, 2008

John Bezzant, *Shepherds and Their Dogs*, Merlin Unwin Books, Ludlow, 2011

K. J. Bonser, *The Drovers: Who They Were and How They Went*, Macmillan, London, 1970

Peter Clery, *Green Gold: A Thousand Years of English Land*, Phillimore & Co., Chichester, 2012

George Culley, *Observations on Live Stock*, London, 1786

H. H. Dixon (The Druid), *Saddle and Sirloin*, Vinton & Co. Ltd, London, 1870

——, *Field and Fern (Scottish Flocks and Herds)*, 2 vols., Vinton & Co., London, 1865

Lord Ernle, *English Farming Past and Present*, 6th edn, Heinemann, London, 1961

Daniel Gates, *Gates's New Shepherd's Guide for Cumberland, Westmoreland and Lancashire*, Brash Bros., Cockermouth, 1879

A. L. J. Gosset, *Shepherds of Britain*, Constable & Co., London, 1911

Stephen Hall and Juliet Clutton-Brock, *Two Hundred Years of British Farm Livestock*, Natural History Museum, London, 1989

Edward Hart, *The Hill Shepherd*, David & Charles, Newton Abbot, 1977

Terry Hearing, *The Dorset Horn: A Short History of the Dorset Horn and Polled Dorset Sheep Breeders' Association*, Dorset Horn and Polled Dorset Sheep Breeders' Association, Dorchester, 1990

Elizabeth Henson, *British Sheep Breeds*, Shire Publications, Oxford, 1986

James Hogg, *Domestic Manners and Private Life of Sir Walter Scott*, Oliver and Boyd, Edinburgh, 1834

——, *The Mountain Bard*, Constable & Co., Edinburgh, 1807

Edward Jesse, *Anecdotes of Dogs*, Bell & Daldy, London, 1870

Norman Jones, *Portland Sheep: A Breed with History*, N. Jones, c.1990

Alistair Moffat, *The Borders: A History of the Borders from Earliest Times*, Deerpark Press, Selkirk, 2002

Sydney Moorhouse, *The British Sheepdog*, H. F. & G. Witherby, London, 1950

National Sheep Association, *British Sheep*, 9th edn, National Sheep Association, Malvern, 1998

Andrew O'Hagan, *The End of British Farming*, Profile Books, London, 2001

Anne Orde (ed.), *Matthew and George Culley, Travel Journals and Letters 1765–1798*, Oxford University Press, Oxford, 2002

Henry Cecil Pawson, *Robert Bakewell: Pioneer Livestock Breeder*, Crosby Lockwood & Sons Ltd, London, 1957

Eileen Power, *The Wool Trade in English Medieval History: Being the Ford Lectures* [1939], Oxford University Press, London, 1941

F. Rainsford-Hannay, *Dry Stone Walling*, Faber & Faber, London, 1957

Ian Roberts, Richard Carlton and Alan Rushworth, *Drove Roads of Northumberland*, History Press, Stroud, 2010

Michael J. H. Robson, *Sheep of the Borders*, Michael J. H. Robson, Newcastleton, 1987

M. L. Ryder, *Sheep and Man*, Duckworth, London, 1983

J. B. Skinner et al. (eds.), *British Sheep and Wool*, British Wool Marketing Board, Bradford, 2010

Henry Tegner, *Charm of the Cheviots*, Frank Graham, Newcastle upon Tyne, 1970

J. F. H. Thomas, *Sheep*, Faber & Faber, London, 1945

Robert Trow-Smith, *A History of British Livestock Husbandry to 1700*, Routledge and Paul, London, 1957

——, *A History of British Livestock Husbandry 1700–1900*, Routledge and Paul, London, 1959

——, *Life from the Land: The Growth of Farming in Western Europe*, Longmans, London, 1967

——, *Society and the Land*, Cresset Press, London, 1953

Peter A. Tulloch, *A Window on North Ronaldsay*, Kirkwall Press, Kirkwall, 1974

Peter Wade-Martins, *Black Faces: A History of East Anglian Sheep Breeds*, Norfolk Museums Service / Geerings, Ashford, Kent, 1993

Rebecca West, *Black Lamb and Grey Falcon: A Journey through Yugoslavia*, Macmillan & Co., London, 1942

Gordon Wilyman, *Memoirs of a Welsh Halfbred*, Gordon Wilyman, 2004

William Youatt, *Sheep: Their Breeds, Management and Diseases*, Baldwin & Cradock, London, 1837

∾ INDEX ∾